JUMP Math

Workbook 6: Part 2

Contents

jump math

MULTIPLYING POTENTIAL.

JUMP Math
Toronto, Ontario
www.jumpmath.org

Writers: Dr. John Mighton, Dr. Sindi Sabourin, Dr. Anna Klebanov
Consultant: Jennifer Wyatt
Cover Design: Blakeley
Special thanks to the design and layout team.
Cover Photograph: © iStockphoto.com/Grafissimo

ISBN: 978-1-897120-49-1

This book is printed on 100% post-consumer waste Forest Stewardship Certified recycled paper, using plant-based inks. The paper is processed chlorine-free and manufactured using biogas energy.

Printed and bound in Canada

A note to educators, parents, and everyone who believes that numeracy is as important as literacy for a fully functioning society

Welcome to JUMP Math

Entering the world of JUMP Math means believing that every child has the capacity to be fully numerate and to love math. Founder and mathematician John Mighton has used this premise to develop his innovative teaching method. The resulting materials isolate and describe concepts so clearly and incrementally that everyone can understand them.

JUMP Math is comprised of workbooks, teacher's guides, evaluation materials, outreach programs, tutoring support through schools and community organizations, and provincial curriculum correlations. All of this is presented on the JUMP Math website: **www.jumpmath.org**.

Teacher's guides are available on the website for free use. Read the introduction to the teacher's guides before you begin using these materials. This will ensure that you understand both the philosophy and the methodology of JUMP Math. The workbooks are designed for use by children, with adult guidance. Each child will have unique needs and it is important to provide the child with the appropriate support and encouragement as he or she works through the material.

Allow children to discover the concepts on the worksheets by themselves as much as possible. Mathematical discoveries can be made in small, incremental steps. The discovery of a new step is like untangling the parts of a puzzle. It is exciting and rewarding.

Children will need to answer the questions marked with a in a notebook. Grid paper and notebooks should always be on hand for answering extra questions or when additional room for calculation is needed. Grid paper is also available in the BLM section of the Teacher's Guide.

The means "Stop! Assess understanding and explain new concepts before proceeding."

1. Write the correct symbol (**+** or **x**) in the circle to make the equation true.

   a) 5 (+) 2 = 7   b) 4 ◯ 1 = 4.   c) 5 ◯ 3 = 8   d) 3 ◯ 5 = 15

   e) 9 ◯ 1 = 10   f) 8 ◯ 5 = 13   g) 2 ◯ 4 = 8   h) 8 ◯ 4 = 12

   i) 2 ◯ 4 = 8   j) 8 ◯ 1 = 9   k) 7 ◯ 3 = 10   l) 7 ◯ 1 = 7

2. Write the correct symbol (**+**, **−**, or **x**) in the circle to make the equation true.

   a) 7 ◯ 3 = 21   b) 2 ◯ 3 = 6   c) 3 ◯ 3 = 9   d) 7 ◯ 1 = 6

   e) 4 ◯ 4 = 8   f) 4 ◯ 4 = 16   g) 9 ◯ 3 = 6   h) 9 ◯ 3 = 12

   i) 9 ◯ 5 = 14   j) 8 ◯ 1 = 9   k) 9 ◯ 1 = 9   l) 3 ◯ 14 = 17

3. Continue the following sequences by **multiplying** each term by the given number.

   a) 3 (x 3) 9 , _____ , _____ , _____   b) 1 (x 3) 3 , _____ , _____ , _____

   c) 4 (x 2) 8 , _____ , _____ , _____   d) 1 (x 7) 7 , _____ , _____ , _____

4. Each term in the sequence below was made by **multiplying** the previous term by a fixed number. Find the number and continue the sequence.

   a) 2 (x) 8 , 32 , _____ , _____   b) 3 (x) 6 , 12 , _____ , _____

   c) 1 (x) 5 , 25 , _____ , _____   d) 2 (x) 10 , 50 , _____ , _____

5. Each of the sequences below was made by **multiplication**, **addition**, or **subtraction**. Continue the sequence.

   a) 1 , 2 , 4 , _____ , _____   b) 5 , 8 , 11 , _____ , _____   c) 18 , 14 ,10 , _____ , _____

   d) 3 , 6 , 12 , _____ , _____   e) 14 , 18 , 22 , _____ , _____   f) 1 , 3 , 9 , _____ , _____

6. Write a rule for each sequence in Question 5.
    (The rule for the first sequence is: "Start at 1, multiply by 2.")

PA6-23: Patterns with Increasing & Decreasing Steps — Part II

1. In the sequences below, the step or gap between the numbers increases or decreases. Can you see a pattern in the way the gap changes?

   Use the pattern to extend the sequence.

   a) 2 , 4 , 7 , 11 , ____ , ____

   b) 3 , 4 , 6 , 9 , 13 , ____ , ____

   c) 12 , 15 , 20 , 27 , ____ , ____

   d) 6 , 8 , 12 , 18 , 26 , ____ , ____

   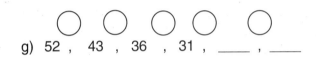

   e) 18 , 13 , 9 , 6 , ____ , ____

   f) 42 , 32 , 24 , 18 , ____ , ____

   g) 52 , 43 , 36 , 31 , ____ , ____

   h) 210 , 180 , 155 , 135 , 120 , ____ , ____

2. Complete the T-table for Figure 3 and Figure 4. Then use the pattern in the gap to predict the number of triangles needed for Figures 5 and 6.

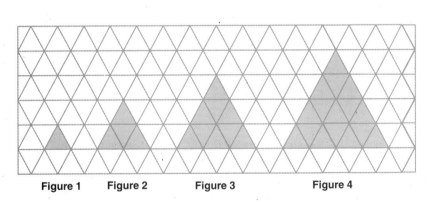

Figure 1   Figure 2   Figure 3   Figure 4

Figure	Number of Triangles	
1	1	◯
2	4	◯
3		◯
4		◯
5		◯
6		

Write the number of triangles added each time here.

3. Make a T-table to predict how many blocks will be needed for Figure 6.

   HINT: Don't forget to count the hidden blocks.

  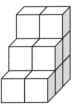

Figure 1   Figure 2   Figure 3

4. In each sequence below, the **step** changes in a regular way (it increases, decreases, or increases and decreases). Write a rule for each pattern.

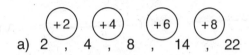

a) 2 , 4 , 8 , 14 , 22

   Rule: Start at 2. Add 2, 4, 6 ... (the step increases by 2).

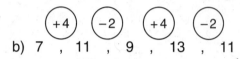

b) 7 , 11 , 9 , 13 , 11

   Rule : Start at 7. Add 4, then subtract 2. Repeat.

c) 2 , 3 , 5 , 8 , 12

   Rule: _____

d) 5 , 7 , 4 , 6 , 3

   Rule: _____

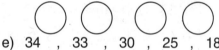

e) 34 , 33 , 30 , 25 , 18

   Rule: _____

5. Write a rule for each pattern.

   HINT: Two of these patterns were made by increasing the step in a regular way, and two were made by multiplication.

   a) 2 , 5 , 10 , 17   b) 2 , 4 , 8 , 16   c) 1 , 3 , 9 , 27   d) 4 , 6 , 10 , 16

6. Write the number of shaded squares or triangles in each figure. Write a rule for the pattern. Use your rule to predict the number of shaded parts in the 5ᵗʰ figure.

   HINT: To count the number of triangles in the last figure in b), try skip counting by 3s.

   a)

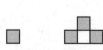

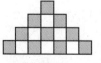

   Figure 1   Figure 2   Figure 3   Figure 4

   b)

   Figure 1   Figure 2   Figure 3   Figure 4

7. Create a pattern with a step that increases and decreases.

PA6-24: Advanced Patterns

Answer the questions below in your notebook.

1. The dots show how many people can sit along each side of a table.

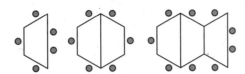

Number of Tables	Number of People

a) Draw a picture to show how many people could sit at 4 and 5 tables. Then, fill in the T-table.

b) Describe the pattern in the number of people. How does the step change?

c) Extend the pattern to find out how many people could sit at 8 tables.

2. a) The Ancient Greeks investigated numbers that could be arranged in geometric shapes.

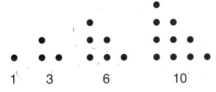

1   3   6   10

The first four **triangular** numbers are shown here to the left.

   i) Find the 5th and 6th triangular numbers by drawing a picture.

   ii) Describe the pattern in the triangular numbers. How does the step change?

   iii) Find the 8th triangular number by extending the pattern you found in ii).

b) Repeat steps i) to iii) with the **square** numbers.

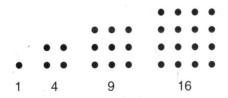

1   4   9   16

3. One of the most famous sequences in mathematics is the **Fibonacci sequence**, shown below.

a) Find the gap between terms, then use the pattern in the gap to continue the sequence.

0   1   3   6   11   19   31   52

1 , 1 , 2 , 3 , 5 , 8 , 13 , 21 , _____ , _____

b) What pattern do you see in the number of odd and even numbers in the Fibonacci sequence?

c) Sum the first 4 odd Fibonacci numbers. Then sum the first 2 even numbers. What do you notice?

d) Sum the first six odd Fibonacci numbers and the first three even numbers. What to you notice?

 jump math
MULTIPLYING POTENTIAL.

Patterns & Algebra 2

PA6-25: Creating and Extending Patterns

1. Extend each pattern for the next three terms. Then write a rule for the pattern.

   a) 237 , 243 , 249 , 255 , 261 , _____ , _____ , _____

      RULE: _____

   b) 6 , 10 , 7 , 11 , 8 , 12 , _____ , _____ , _____

      RULE: _____

   c) 47 , 45 , 42 , 38 , _____ , _____ , _____

      RULE: _____

2. Use the letters of the alphabet to continue the following patterns.

   A B C D E F G H I J K L M N O P Q R S T U V W X Y Z

   a) A , D , G , J , ____ , ____   b) A , E , J , P , ____

   c) Z , X , V , T , ____ , ____   d) W , X , U , V , ____ , ____

   e) A , C , F , J , O , ____   f) Z , Y , W , T , ____ , ____

   g) A , A , B , A , B , C , ____ , ____ , ____ , ____

3. Create your own pattern using the letters of the alphabet. Write a rule for your pattern.

4. Write out the first five terms of each pattern.
   a) Start at 3. Multiply by 3 each time.
   b) Start at 39. Add 3 to the first number, 5 to the second number, 7 to the third number and so on.
   c) Start at 375. Repeatedly subtract 4 and add 3.
   d) Start at 1. Multiply by 2 and add 1 each time.

5. Create a pattern for each condition.
   a) The numbers increase, then decrease, then increase, then decrease, and so on.
   b) The pattern increases by multiplying each term by the same number.

6. If A = 1, B = 2, C = 3, and so on, what is the value of E × Y?

PA6-26: Patterns with Larger Numbers

1. Use addition or multiplication to complete the following charts.

a)

Years	Weeks
1	52
2	
3	
4	

b)

Years	Days
1	365
2	
3	

c)

Hours	Seconds
1	3600
2	
3	
4	

2. Water drains from two tanks at the rate shown in the chart below. Describe the pattern in the numbers in each column.

   Which tank do you think will empty first?

Minutes	Tank 1	Tank 2
1	500 L	500 L
2	460 L	490 L
3	420 L	470 L
4	380 L	440 L

3. a) How much fuel will be left in the airplane after 25 minutes?

   b) How far from the airport will the plane be after 30 minutes?

   c) How much fuel will be left in the airplane when it reaches the airport?

Minutes	Litres of Fuel	km from Airport
0	1200	525
5	1150	450
10	1100	375

4.  Halley's Comet returns to Earth every 76 years. It was last seen in 1986.

   a) How many times will the comet be seen in the 2000s?

   b) When was the first time Halley's Comet was seen in the 1900s?

5. Use multiplication to find the first few products. Look for a pattern.
   Use the pattern you've described to fill in the rest of the numbers.

   a) $37 \times 3 = $ _____

   $37 \times 6 = $ _____

   $37 \times 9 = $ _____

   $37 \times 12 = $ _____

   _____ $ = $ _____

   b) $9 \times 2222 = $ _____

   $9 \times 3333 = $ _____

   $9 \times 4444 = $ _____

   $9 \times 5555 = $ _____

   _____ $ = $ _____

6. Using a calculator, can you discover any patterns like the ones in Question 5?

jump math
MULTIPLYING POTENTIAL

Patterns & Algebra 2

PA6-27: Equations

1. In a word problem an empty box can stand for an unknown quantity.
   Find the missing number in each problem and write it in the box.

a) There are 10 marbles   4 are outside the box   How many are inside?

$$10 = 4 + \boxed{}$$

b) There are 9 marbles   6 are outside the box   How many are inside?

$$9 = 6 + \boxed{}$$

c) There are 12 children in a class   7 are girls   How many are boys?

$$12 = 7 + \boxed{}$$

d) A cat had 7 kittens   4 kittens are boys   How many are girls?

$$7 = 4 + \boxed{}$$

e) Paul had some stickers   He gave away 3   4 were left

$$\boxed{} - 3 = 4$$

f) There are 15 oranges in boxes   How many oranges are in each box?   There are 3 boxes

$$15 \div \boxed{} = 3$$

2. Find the number that makes each equation true (by guessing and checking) and write it in the box.

a) $\boxed{} + 4 = 7$

b) $\boxed{} + 3 = 6$

c) $\boxed{} + 5 = 9$

d) $9 - \boxed{} = 6$

e) $17 - \boxed{} = 13$

f) $11 - \boxed{} = 9$

g) $2 \times \boxed{} = 6$

h) $5 \times \boxed{} = 15$

i) $3 \times \boxed{} = 9$

j) $\boxed{} \div 2 = 4$

k) $\boxed{} \div 5 = 3$

l) $\boxed{} \div 3 = 4$

m) $5 + 4 = 6 + \boxed{}$

n) $10 - 4 = \boxed{} + 5$

o) $\boxed{} + \boxed{} + 2 = 8$

3. Find two different answers for the following equation:

$$\boxed{} + \boxed{} + \bigcirc = 5 \qquad \boxed{} + \boxed{} + \bigcirc = 5$$

4. How many answers can you find for the equation: $\boxed{} + \boxed{} + \bigcirc = 9$?

PA6-28: Equations (Advanced)

1. What number does the letter represent?

a) $x + 3 = 9$

$x = \boxed{}$

b) $A - 3 = 5$

$A = \boxed{}$

c) $n + 5 = 11$

$n = \boxed{}$

d) $6x = 18$

$x = \boxed{}$

e) $y + 5 = 17$

$y = \boxed{}$

f) $3n = 15$

$n = \boxed{}$

g). $b \div 2 = 8$

$b = \boxed{}$

h) $4x = 20$

$x = \boxed{}$

i) $z - 2 = 23$

$z = \boxed{}$

j) $m - 2 = 25$

$m = \boxed{}$

2. What number does the box or the letter "n" represent? (Guess and check.)

a) $2 \times \boxed{3} + 3 = 9$

$\square = \boxed{}$

b) $5 \times \boxed{1} - 2 = 8$

$\square = \boxed{}$

c) $3 \times \boxed{2} + 5 = 14$

$\square = \boxed{}$

d) $2 \times \boxed{1} - 5 = 3$

$\square = \boxed{}$

e) $7 \times \boxed{2} + 2 = 16$

$\square = \boxed{}$

f) $n + 5 = 4 + 10$

$n = \boxed{}$

g) $n - 2 = 12 - 4$

$n = \boxed{}$

h) $4n + 1 = 13$

$n = \boxed{}$

i) $5n + 2 = 27$

$n = \boxed{}$

3. Find x.

a) $x + x = 8$

$x = \underline{}$

b) $x + x + x = 12$

$x = \underline{}$

c) $x + x + x = 24$

$x = \underline{}$

4. Find all values of a and b (that are whole numbers) that make the equation true.

a) $a + b = 6$

b) $a \times b = 6$

c) $6 - a = b$

5. If $2a + 6 = 12$ and $2b + 6 = 14$, explain why b must be greater than a.

6. Write 3 different equations with solution 5.

7.

A	A	A	12
A	B	B	14
A	B	C	10

The shaded column shows the total of each row.

For instance $A + A + A = 12$

Find A, B, and C.

jump math
MULTIPLYING POTENTIAL.

Patterns & Algebra 2

PA6-29: Variables

A **variable** is a letter or symbol (such as **x**, **n**, or **h**) that represents a number.

In the product of a number and a variable, the multiplication sign is usually dropped.

> $3 \times T$ is written $3T$, and $5 \times z$ is written $5z$.

---

1. Write a numerical expression for the cost of renting a kayak …

   a) 2 hours: __5 x 2 = 10__   b) 4 hours: _____   c) 7 hours: _____

Rent a kayak
$5 for each hour

2. Write an expression for the distance a car would travel at ….

   a) Speed: 70 km per hour
   Time: 3 hours

   Distance: _____ km

   b) Speed: 40 km per hour
   Time: 2 hours

   Distance: _____ km

   c) Speed: 100 km per hour
   Time: h hours

   Distance: _____ km

3. Write an algebraic expression for the cost of renting a sailboat for …

   a) h hours: _____ or _____   b) t hours: _____ or _____

   c) x hours: _____ or _____   d) n hours: _____ or _____

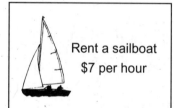

Rent a sailboat
$7 per hour

4. Write an equation that tells you the relationship between the numbers in column A and column B.

   a)

A	B
1	5
2	6
3	7

   __A + 4 = B__

   b)

A	B
1	3
2	6
3	9

   __3 × A = B__

   c)

A	B
1	8
2	9
3	10

   _____

   d)

A	B
1	5
2	10
3	15

   _____

   e)

A	B
1	8
2	16
3	24

   _____

5. Replace the letter in x in x + 5 = 9 with any other letter.
   Does the equation still give the same solution?
   Explain.

6. Write an equation for each problem. (Use a variable for the unknown)

   a) In a class of 28 students, 15 are girls.
   How many students are boys?

   b) Ramona had 48 stamps but gave some away.
   She kept 24. How many did she give away?

PA6-30: Algebraic Puzzles

1. Scale A is balanced perfectly. Draw the number of circles needed to balance Scale B.

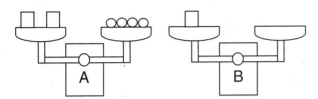

2. Scale A is balanced perfectly. Draw the number of circles needed to balance Scale B.

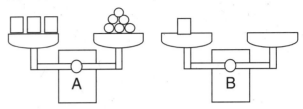

3. Scale A is balanced perfectly. Draw the number of triangles needed to balance Scale B.

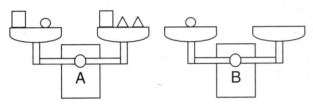

4. Scale A is balanced perfectly. Draw the number of circles needed to balance Scale B.

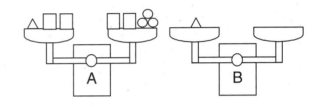

BONUS

5. Scales A and B are balanced perfectly. Draw the number of circles needed to balance Scale C.

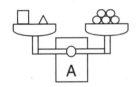 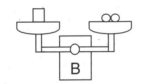 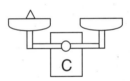

6. Scales A and B are balanced perfectly. Draw the number of circles needed to balance Scale C.

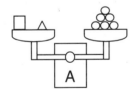 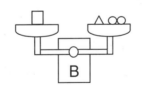 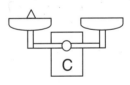

7. Fill in the missing digits.

a)
```
    4 8
 + 1 ☐
 ─────
   6 2
```

b)
```
   3 ☐
 + 2 7
 ─────
   6 4
```

c)
```
   8 1
 - 3 ☐
 ─────
   4 8
```

d)
```
   6 3
 - ☐ 9
 ─────
   2 4
```

e)
```
   3 ☐
 ×   4
 ─────
 1 2 8
```

f)
```
   5 ☐
 ×   3
 ─────
 1 6 8
```

g)
```
   2 3
 ×   ☐
 ─────
   9 2
```

h)
```
   8 3 4 5
 - 2 ☐ 7 ☐
 ─────────
   ☐ 4 ☐ 7
```

8. Explain how you found your answer in Question 5.

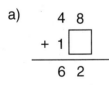 jump math
MULTIPLYING POTENTIAL.

Patterns & Algebra 2

PA6-31: Graphs

1. For each set of points, write a list of ordered pairs, and then complete the T-table.

a)
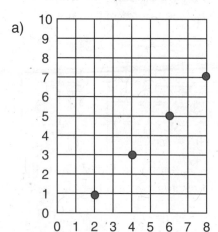

b)
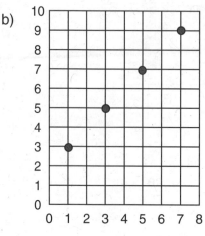

c)
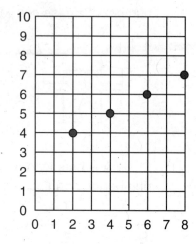

Ordered Pairs	First Number	Second Number
(2 , 1)	2	1
(,)		
(,)		
(,)		

Ordered Pairs	First Number	Second Number
(,)		
(,)		
(,)		
(,)		

Ordered Pairs	First Number	Second Number
(,)		
(,)		
(,)		
(,)		

2. Mark four points on the line segments. Then write a list of ordered pairs, and complete the T-table.

a)
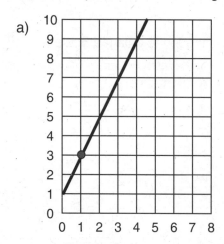

b)
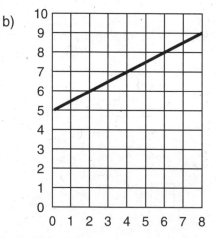

c)

Ordered Pairs	First Number	Second Number
(1 , 3)	1	3
(,)		
(,)		
(,)		

Ordered Pairs	First Number	Second Number
(,)		
(,)		
(,)		
(,)		

Ordered Pairs	First Number	Second Number
(,)		
(,)		
(,)		
(,)		

Patterns & Algebra 2

PA6-31: Graphs (continued)

3. Write a list of ordered pairs based on the T-table provided. Mark the ordered pairs on the graph and connect the points to form a line.

First Number	Second Number
3	1
4	3
5	5
6	7

(,)

(,)

(,)

(,)

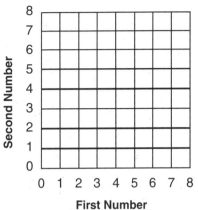

4. Draw a graph for each T-table (as in Question 1).
   NOTE: Make sure you look carefully at the scale in part d).

a)

Input	Output
2	5
4	6
6	7
8	8

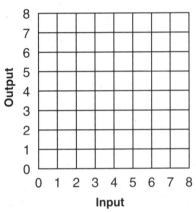

b)

Input	Output
1	7
2	6
3	5
4	4

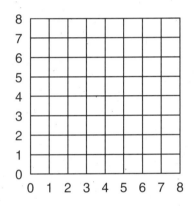

BONUS

c)

Input	Output
2	4
4	8
6	12
8	16

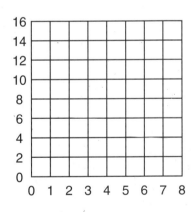

d)

Input	Output
1	6
3	8
5	10
7	12

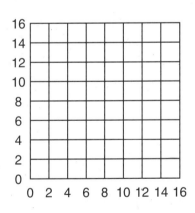

5. Draw a coordinate grid (like those above) on grid paper and plot the following ordered pairs:
   (1 , 2), (3 , 5), (5 , 8), and (7 , 11).

6. On grid paper, make a T-table and graph for the following rules.

   a) Multiply by 2 and subtract 1

   b) Multiply by 4 and subtract 3

   c) Divide by 2 and add 3

7. Make a T-table for each set of points on the coordinate grid.
   Write a rule for each T-table that tells you how to calculate the output from the input.
   (See the rules in Questions 6.)

Graph A

Input	Output

Graph B

Input	Output

Graph C

Input	Output

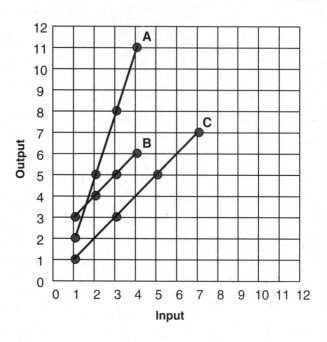

Rule for T-table A: _____

Rule for T-table B: _____

Rule for T-table C: _____

8. Mark <u>four</u> points that lie on a straight line in the coordinate grid. Then, make a T-table for your set of points.

First Number	Second Number

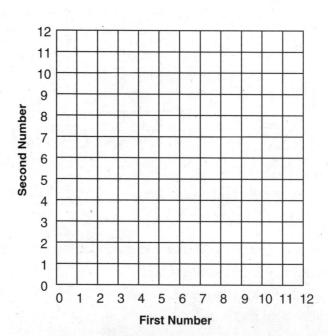

Answer these questions in your notebook.

1.

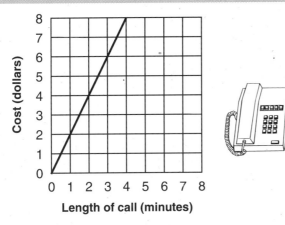

Length of call (minutes)

The graph shows the cost of making a long distance telephone call.

a) If you talked for 2 minutes, how much would you pay?

b) What is the cost for a minute call?

c) How much would you pay to talk for 10 minutes?

d) If you paid 6 dollars, how long would you be able to talk for?

e) How much would you pay to talk for 30 seconds?

2.

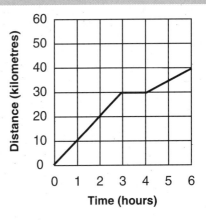

Time (hours)

The graph shows the distance Kathy travelled on a cycling trip.

a) How far had Kathy cycled after 2 hours?

b) How far had Kathy travelled after 6 hours?

c) Did Kathy rest at all on her trip? How do you know?

d) When she was cycling, did Kathy always travel at the same speed?

3.

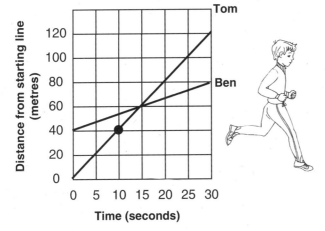

Time (seconds)

Ben and Tom ran a 120 m race.

a) How far from the start was Tom after 10 seconds?

b) How far from the start was Ben after 15 seconds?

c) Who won the race? By how much?

d) How much of a head start did Ben have?

e) How many seconds from the start did Tom overtake Ben?

4.

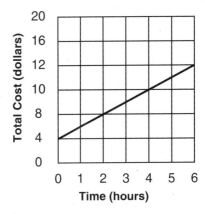

Time (hours)

The graph shows the cost of renting a bike from Mike's store.

a) How much would you pay to rent the bike for:
   i) 2 hours?   ii) 4 hours?   iii) 3 hours?

b) How much do you pay for the bike before you have even used it?

c) Dave's store charges $3.50 an hour for a bike. Whose store would you rent from if you wanted the bike for 3 hours? Explain.

1. For each T-table below, write a rule that tells:
   - how the input changes,
   - how the output changes, and
   - the relation between the input and output.

Example:
   - The numbers in the input column increase by 1 each time.
   - The numbers in the output column increase by 3 each time.
   - Multiply the input by 3 and add 2 to get the output.

INPUT	OUTPUT
1	5
2	8
3	11
4	14

a)

INPUT	OUTPUT
1	6
2	9
3	12

b)

INPUT	OUTPUT
1	21
2	22
3	23

c)

INPUT	OUTPUT
1	6
2	10
3	14

d)

INPUT	OUTPUT
1	7
2	14
3	21

e)

INPUT	2.5	3.0	4.0	5.5	7.5	10.0	13.0
OUTPUT	5	6	8	11	15	20	26

f)

INPUT	1	2	3	4	5	6	7
OUTPUT	1	4	9	16	25	36	49

g)

INPUT	1	2	3	4	5	6	7
OUTPUT	2.1	4.2	6.3	8.4	10.5	12.6	14.7

2. The chart shows the number of kilometres Karen can run in 15 minutes.

   Complete the chart.

   NOTE: Assume she keeps running at the same rate.

Distance	Time (seconds)	Time (minutes)	Time (hours)
2.3 km		15	$\frac{1}{4}$
4.6 km			

Investigation

There are eight dots in Figure 1.

Each pair of dots is joined by exactly 1 line segment (●——●).

How can you find out how many line segments there are without counting every line?

A mathematician would start with fewer dots and use a pattern to make a prediction.

Figure 1

1. For each set of dots below, use a ruler to join every pair of dots with a straight line. Write the number of lines in the space provided.

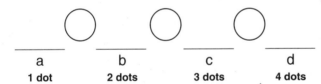

a)   b)   c)   d)

1 dot   2 dots   3 dots   4 dots

_____ lines   _____ lines   _____ lines   _____ lines

TEACHER: Check that every student has the correct answer to Question 1 before moving on.

2. Write the numbers you found in Question 1 on the lines above their appropriate letter (a, b, c or d). Find the gaps between the numbers and write your answers in the circles provided.

a   b   c   d
1 dot   2 dots   3 dots   4 dots

3. Predict the gaps and numbers in the sequence. Write your predictions below.

○   ○   ○   ○   ○

_____   _____   _____   _____   _____   _____
1 dot   2 dots   3 dots   4 dots   5 dots   6 dots

4. Test your predictions by joining the dots in each figure.
   Were you right?

5 dots   6 dots

5. a) Use the rule you discovered to calculate the number of line segments in Figure 1.

_____

   b) If you joined every pair of dots in a set of 10 dots, how many lines would you need?

_____

1. The picture below shows how many chairs can be placed at each arrangement of tables.

   a) Make a T-table and state a rule that tells the relationship between the number of tables and the number of chairs.

   b) How many chairs can be placed at 12 tables?

2. Andy has $10 in his bank account.
   He saves 25 dollars each month.
   How much does he have in his account after 10 months?

3. A recipe calls for 5 cups of flour for every 6 cups of water.
   How many cups of water will be needed for 25 cups of flour?

4. Raymond is 400 km from home on Wednesday morning.
   He cycles 65 km toward home each day.
   How far away from home is he by Saturday evening?

5. Every 6th person who arrives at a book sale receives a free calendar.
   Every 8th person receives a free book.
   Which of the first 50 people receive a book and a calendar?

6. Anna's basket holds 24 apples and Emily's basket holds 30 apples.
   They each collected less than 150 apples.
   How many baskets did they collect if they collected the same number of apples?

7. a) How many shaded squares will be on the perimeter of the 10th figure? How do you know?

   b) How many white squares will be in a figure that has a shaded perimeter of 32 squares?

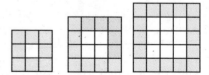

8. Gerome wants to rent a hockey rink for 6 hours. Which is the cheapest way to rent the rink:
   (i) pay a fee of $60 for the first hour and $35 for each hour after that?   or   (ii) pay $45 each hour?

PA6-35: Problems and Puzzles (continued)

9. What strategy would you use to find the 72nd shape in this pattern? What is the shape?

10. Paul shovelled sidewalks for 4 days.
    Each day, he shovelled 3 more sidewalks than the day before.
    He shovelled 30 sidewalks altogether.
    How many sidewalks did he shovel on each of the days? **Guess and check!**

11. Make a chart with three columns to show:
    - the number of edges along a side of the figure,
    - the number of small triangles in the figure,
    - the perimeter of the figure.

    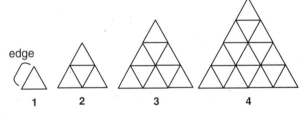

    Describe the pattern in each column and any relationships between the columns of the chart.

12. The picture shows how the temperature changes at different heights over a mountain.

    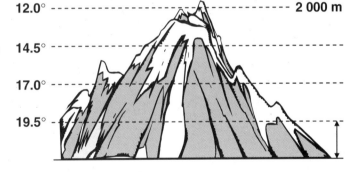

    a) Does the temperature increase or decrease at greater heights?

    b) What distance does the arrow represent in real life?

    c) Measure the length of the arrow. What is the scale of the picture?

        _____ cm = _____ m

    d) Do the numbers in the sequence of temperatures decrease by the same amount each time?

    e) If the pattern in the temperature continued, what would the temperature be at:

        (i) 3000 m?   (ii) 4000 m?

13. Marlene says she will need 27 blocks to make Figure 7.
    Is she right?
    Explain.

        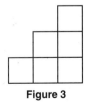

    Figure 1   Figure 2   Figure 3

Fractions name equal parts of a whole.

The pie is cut into 4 equal parts. 3 parts out of 4 are shaded.

$\frac{3}{4}$ of a pie is shaded.

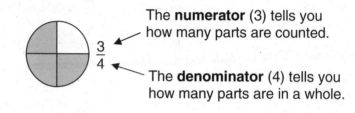

The **numerator** (3) tells you how many parts are counted.

$\frac{3}{4}$

The **denominator** (4) tells you how many parts are in a whole.

----

1. Name the following fractions.

   a)   b)   c) 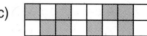   d)

2. Use a **ruler** to divide each box into equal parts.

   a) 3 equal parts   b) 10 equal parts

      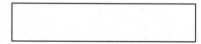

3. Using a **ruler**, find what fraction of each of the following boxes is shaded.

   a)    b)

   _____ is shaded.   _____ is shaded.

4. Using a ruler, complete the following figures to make a whole.

   a) $\frac{1}{4}$ b) 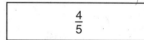 $\frac{1}{2}$ c) $\frac{4}{5}$

5. You have $\frac{5}{8}$ of a pie.

   a) What does the bottom (denominator) of the fraction tell you?

   b) What does the top (numerator) of the fraction tell you?

6. In your notebook explain why each picture does or does not show $\frac{1}{4}$.

   a)   b)   c)   d)

7.   Draw three 4 × 4 grids on grid paper.
   Show three different ways to shade half of the grid.
   HINT: The picture shows one way.

jump math
MULTIPLYING POTENTIAL.

Number Sense 2

Fractions can name parts of a set: $\frac{3}{5}$ of the figures are pentagons, $\frac{1}{5}$ are squares and $\frac{1}{5}$ are circles.

----

1. Fill in the blanks.

   a)

   _____ of the figures are pentagons.

   _____ of the figures are shaded.

   b)

   _____ of the figures are squares.

   _____ of the figures are shaded.

2. Fill in the blanks.

   a) $\frac{4}{7}$ of the figures are _____

   b) $\frac{2}{7}$ of the figures are _____

   c) $\frac{1}{7}$ of the figures are _____

   d) $\frac{5}{7}$ of the figures are _____

3.    Describe this picture in two different ways using the fraction $\frac{3}{5}$.

4. A hockey team wins 6 games, loses 4 games, and ties one game.
   What fraction of the games did the team …

   a) win? _____   b) lose? _____   c) tie? _____

5. A box contains 2 blue marbles, 3 red marbles, and 4 yellow marbles.

   What fraction of the marbles are **not** blue? _____

6. The chart shows the number of children with each given hair colour in a class.

Hair Colour	Black	Brown	Red	Blonde
Number of Children	5	5	1	3

What fraction of children in the class has hair that is …

a) red? b) black? c) blonde? d) brown?

7. There are 23 children in a class.
Each child chose to do a science project on animals or on plants.
The chart shows the number who chose each topic.

a) Fill in the missing numbers in the chart.

b) What fraction of the children chose to study …

    animals?   plants?

c) What fraction of the girls chose to study …

    animals?   plants?

	Animals	Plants
Boys	7	4
Girls		
Children	12	

8. What fraction of the **squares** are on the **outside** of the figure? _____

9. Write a **fraction** for each statement below.

 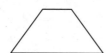

a) ☐ of the figures are pentagons.

b) ☐ of the figures have 4 vertices.

c) ☐ of the figures have exactly 2 right angles.

d) ☐ of the figures have exactly 1 pair of parallel sides.

10. Write two fractions statements for the figures in Question 9 above. Justify your answer.

1.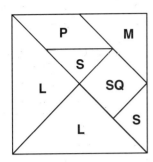

In a tangram ...

- 2 small triangles (**S**) cover a medium triangle (**M**)
- 2 small triangles (**S**) cover a square (**SQ**)
- 2 small triangles (**S**) cover a parallelogram (**P**)
- 4 small triangles (**S**) cover a large triangle (**L**)

What fraction of each shape is covered by a <u>single</u> small triangle?

a)

b)

c)

d)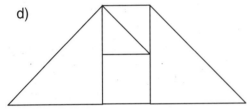

e)

f)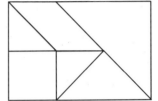

2. What fraction of each shape is shaded? Explain how you know.

a)

b)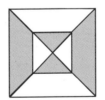

c)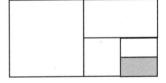

3. What fraction of the trapezoid is covered by a <u>single</u> small triangle?

Show your work.

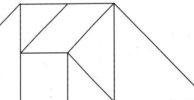

4. If 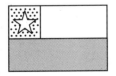 = red and = blue, approximately what fraction of each flag is shaded red? Explain.

a)

CHILE

b)

CANADA

c)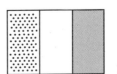

FRANCE

d)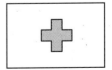

SWITZERLAND

Number Sense 2

1. Imagine moving the shaded pieces from pies A and B into pie plate C. Show how much of pie C would be filled then write a fraction for pie C.

A 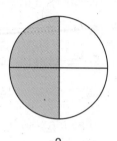 **B**   **C**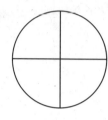

$$\frac{1}{4} \quad + \quad \frac{2}{4} \quad = \quad \underline{\quad}$$

2. Imagine pouring the liquid from cups A and B into cup C.
   Shade the amount of liquid that would be in C.
   Then complete the addition statements.

   a) 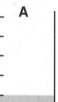   b)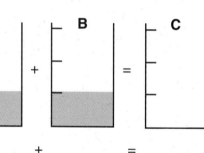

$$\frac{\quad}{5} \quad + \quad \frac{\quad}{5} \quad = \quad \underline{\quad} \qquad\qquad \frac{\quad}{3} \quad + \quad \frac{\quad}{3} \quad = \quad \underline{\quad}$$

3. Add.

   a) $\frac{3}{5} + \frac{1}{5} =$ b) $\frac{2}{4} + \frac{1}{4} =$ c) $\frac{3}{7} + \frac{2}{7} =$ d) $\frac{5}{8} + \frac{2}{8} =$

   e) $\frac{3}{11} + \frac{7}{11} =$ f) $\frac{5}{17} + \frac{9}{17} =$ g) $\frac{11}{24} + \frac{10}{24} =$ h) $\frac{18}{57} + \frac{13}{57} =$

4. Show how much pie would be left if you took away the amount shown. Then complete the fraction statement.

   a) 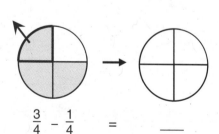   b)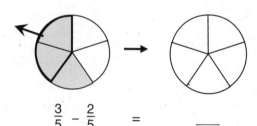

$$\frac{3}{4} - \frac{1}{4} \quad = \quad \underline{\quad} \qquad\qquad\qquad \frac{3}{5} - \frac{2}{5} \quad = \quad \underline{\quad}$$

5. Subtract.

   a) $\frac{2}{3} - \frac{1}{3} =$ b) $\frac{3}{5} - \frac{2}{5} =$ c) $\frac{6}{7} - \frac{3}{7} =$ d) $\frac{5}{8} - \frac{2}{8} =$

   e) $\frac{9}{12} - \frac{2}{12} =$ f) $\frac{6}{19} - \frac{4}{19} =$ g) $\frac{9}{28} - \frac{3}{28} =$ h) $\frac{17}{57} - \frac{12}{57} =$

NS6-58: Ordering and Comparing Fractions

1.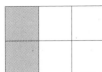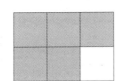

   What fraction has a greater numerator, $\frac{2}{6}$ or $\frac{5}{6}$? _____

   Which fraction is greater? _____

   Explain your thinking: _____

   _____

2. Circle the greater fraction in each pair.

   a) $\frac{5}{17}$ or $\frac{11}{17}$
   b) $\frac{3}{17}$ or $\frac{4}{17}$
   c) $\frac{11}{25}$ or $\frac{6}{25}$
   d) $\frac{57}{115}$ or $\frac{43}{115}$

3. Two fractions have the same <u>denominators</u> (bottoms) but different <u>numerators</u> (tops). How can you tell which fraction is greater?

   _____

4. Write the fractions in order from least to greatest.

   a) $\frac{4}{5}$, $\frac{1}{5}$, $\frac{3}{5}$

   b) $\frac{9}{10}$, $\frac{2}{10}$, $\frac{1}{10}$, $\frac{5}{10}$

5. Circle the greater fraction in each pair.

   a) $\frac{1}{7}$ or $\frac{1}{6}$
   b) $\frac{8}{8}$ or $\frac{8}{9}$
   c) $\frac{7}{300}$ or $\frac{7}{200}$

6. Fraction A and Fraction B have the same <u>numerators</u> (tops) but different <u>denominators</u> (bottoms). How can you tell which fraction is greater?

   _____

7. Write the fractions in order from least to greatest.

   a) $\frac{1}{9}$, $\frac{1}{4}$, $\frac{1}{17}$

   b) $\frac{2}{11}$, $\frac{2}{5}$, $\frac{2}{7}$, $\frac{2}{16}$

8. Circle the greater fraction in each pair.

   a) $\frac{2}{3}$ or $\frac{2}{9}$
   b) $\frac{7}{17}$ or $\frac{11}{17}$
   c) $\frac{6}{288}$ or $\frac{6}{18}$

9. Which fraction is greater, $\frac{1}{2}$ or $\frac{1}{100}$? Explain your thinking.

10. Is it possible for $\frac{2}{3}$ of a pie to be bigger than $\frac{3}{4}$ of another pie? Show your thinking with a picture.

jump math
MULTIPLYING POTENTIAL.

NS6-59: Mixed Fractions

William and Jessie ate three and three quarter pies altogether (or $3\frac{3}{4}$ pies).

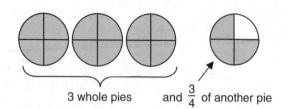

3 whole pies and $\frac{3}{4}$ of another pie

> **NOTE:** $3\frac{3}{4}$ is called a **mixed fraction** because it is a mixture of a whole number and a fraction.

----

1. Write how many <u>whole</u> pies are shaded.

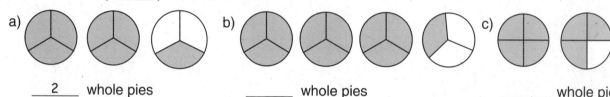

    <u> 2 </u> whole pies   _____ whole pies   _____ whole pie

2. Write each fraction as a <u>mixed fraction</u>.

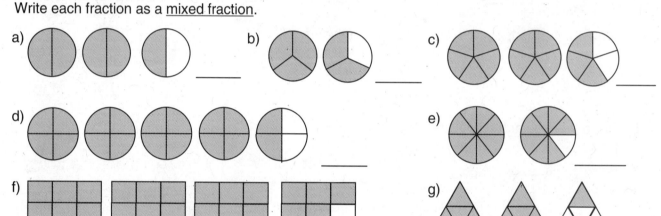

3. Shade the amount of pie given in bold.
   NOTE: There may be more pies than you need.

a) $2\frac{2}{3}$

b) $3\frac{1}{4}$

c) $1\frac{3}{4}$

d) $2\frac{4}{5}$

4. Sketch.   a) $2\frac{1}{4}$ pies   b) $3\frac{2}{3}$ pies   c) $1\frac{1}{5}$ pies   d) $3\frac{1}{6}$ pies

5. Which fraction represents more pie, $3\frac{2}{3}$ or $4\frac{1}{4}$?

6. Is $5\frac{3}{4}$ closer to 5 or 6?

jump math
MULTIPLYING POTENTIAL.

Number Sense 2

NS6-60: Improper Fractions

Huan-Yue and her friends ate **9** quarter-sized pieces of pizza.

 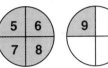

$$\frac{9}{4} = 2\frac{1}{4}$$

improper fraction   mixed fraction

Altogether they ate $\frac{9}{4}$ pizzas.

When the numerator of a fraction is larger than the denominator, the fraction represents more than a whole. Such fractions are called **improper fractions**.

- -

1. Write these fractions as <u>improper</u> fractions.

a)   ____

b)  ____

c)  ____

d)   ____

e)    ____

f) 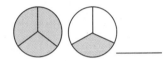 ____

g)  ____

h) 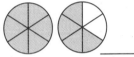 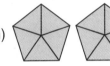   ____

2. Shade one piece at a time until you have shaded the amount of pie given in bold.

a) $\frac{5}{2}$

b) $\frac{9}{4}$

c) $\frac{10}{3}$

d) $\frac{8}{4}$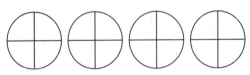

 3. Sketch. a) $\frac{9}{4}$ pies b) $\frac{7}{3}$ pies c) $\frac{9}{2}$ pies d) $\frac{7}{6}$ pies

4. Which fraction represents more pie? $\frac{7}{4}$ or $\frac{9}{4}$? How do you know?

5. Which fractions are more than a whole? How do you know? a) $\frac{5}{7}$ b) $\frac{9}{8}$ c) $\frac{13}{11}$

jump math
MULTIPLYING POTENTIAL.

Number Sense 2

NS6-61: Mixed and Improper Fractions

1. Write these fractions both as <u>mixed</u> fractions and as <u>improper</u> fractions.

   a)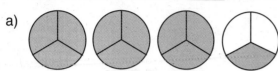

   b)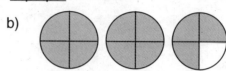

   c)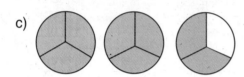

   d)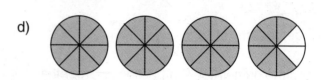

   e)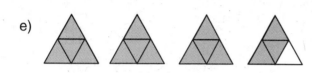

   f)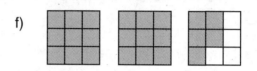

2. Shade the amount of pie given in bold. Then write an <u>improper</u> fraction for the amount of pie.

   a) $3\frac{1}{2}$

   Improper Fraction: _____

   b) $3\frac{3}{4}$

   Improper Fraction: _____

3. Shade the amount of pie given in bold. Then write a <u>mixed</u> fraction for the amount of pie.

   a) $\frac{7}{3}$

   Mixed Fraction: _____

   b) $\frac{19}{6}$

   Mixed Fraction: _____

   c) $\frac{13}{4}$

   Mixed Fraction: _____

   d) $\frac{13}{5}$

   Mixed Fraction: _____

   e) $\frac{25}{8}$

   Mixed Fraction: _____

   f) $\frac{19}{4}$

   Mixed Fraction: _____

 4. Draw a picture to find out which fraction is greater.

   a) $2\frac{1}{2}$ or $\frac{3}{2}$

   b) $2\frac{4}{5}$ or $\frac{12}{5}$

   c) $\frac{15}{8}$ or $\frac{7}{3}$

5. How could you use division to find out how many <u>whole</u> pies are in $\frac{11}{3}$ of a pie? Explain.

Number Sense 2

NS6-62: Mixed Fractions (Advanced)

 There are 4 quarter pieces in 1 pie.

 There are 8 (2 × 4) quarters in 2 pies.

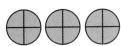 There are 12 (3 × 4) quarters in 3 pies.

How many quarter pieces are in $3\frac{3}{4}$ pies?

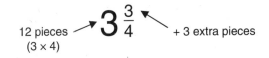

12 pieces (3 × 4) $3\frac{3}{4}$ + 3 extra pieces

So there are 15 quarter pieces altogether.

---

1. Find the number of **halves** in each amount.

   a) 1 pie = _____ halves

   b) 2 pies = _____ halves

   c) 4 pies = _____ halves

   d) $3\frac{1}{2}$ pies = _____ halves

   e) $4\frac{1}{2}$ pies = _____ halves

   f) $5\frac{1}{2}$ pies = _____

2. Find the number of **thirds** in each amount.

   a) 1 pie = _____ thirds

   b) 2 pies = _____ thirds

   c) 4 pies = _____ thirds

   d) $1\frac{1}{3}$ pies = _____ thirds

   e) $2\frac{2}{3}$ pies = _____

   f) $5\frac{2}{3}$ pies = _____

3. A box holds 4 cans.

   a) 2 boxes hold _____ cans

   b) $2\frac{1}{4}$ boxes hold _____ cans

   c) $3\frac{3}{4}$ boxes hold _____ cans

4. A box holds 6 cans.

   a) $2\frac{1}{6}$ boxes hold _____ cans

   b) $2\frac{5}{6}$ boxes hold _____ cans

   c) $3\frac{1}{6}$ boxes hold _____ cans

5. Write the following mixed fractions as improper fractions.

   a) $2\frac{1}{3} = \frac{}{3}$

   b) $5\frac{1}{2} = \frac{}{2}$

   c) $4\frac{2}{3} = \frac{}{3}$

   d) $6\frac{1}{4} = \frac{}{4}$

6. Envelopes come in packs of 6.
   Alice used $2\frac{5}{6}$ packs.
   How many envelopes did she use? _____

7. Baseball cards come in packs of 8. How many cards are in $3\frac{1}{2}$ packs? _____

8. Maia and her friends ate $2\frac{3}{4}$ pizzas. How many quarter-sized pieces did they eat? _____

9.  **A**   Cindy needs $2\frac{2}{3}$ cups of flour.

   $\frac{1}{3}$ cup

   a) How many scoops of cup A would she need? _____

    B

   $\frac{1}{6}$ cup

   b) How many scoops of cup B would she need? _____

Number Sense 2

NS6-63: Mixed & Improper Fractions (Advanced)

How many whole pies are there in $\frac{13}{4}$ pies?

There are 13 pieces altogether, and each pie has 4 pieces.
So you can find the number of whole pies by dividing 13 by 4: $13 \div 4 = 3$ remainder 1

There are 3 whole pies and 1 quarter left over, so: $\frac{13}{4} = 3\frac{1}{4}$

---

1. Find the number of whole pies in each amount by dividing.

   a) $\frac{4}{2}$ pies = _____ whole pies   b) $\frac{6}{2}$ pies = _____ whole pies   c) $\frac{12}{2}$ pies = _____ whole pies

   d) $\frac{6}{3}$ pies = _____ whole pies   e) $\frac{15}{3}$ pies = _____ whole pies   f) $\frac{8}{4}$ pies = _____ whole pies

2. Find the number of whole and the number of pieces remaining by dividing.

   a) $\frac{5}{2}$ pies = ___2___ whole pies and ___1___ half pies = ___$2\frac{1}{2}$___ pies

   b) $\frac{11}{3}$ pies = _____ whole pies and _____ thirds = _____ pies

   c) $\frac{10}{3}$ pies = _____ whole pies and _____ thirds = _____ pies

   d) $\frac{9}{2}$ pies = _____ whole pies and _____ half pies = _____ pies

3. Write the following improper fractions as mixed fractions.

   a) $\frac{3}{2}$ =   b) $\frac{9}{2}$ =   c) $\frac{8}{3}$ =   d) $\frac{15}{4}$ =   e) $\frac{22}{5}$ =

4. Write a mixed and improper fraction for the number of litres.

5. Write a mixed and improper fraction for the length of the rope.

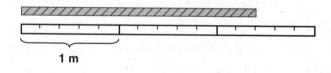

6. Which is greater: $\frac{7}{3}$ or $\frac{5}{2}$? How do you know?

7. Between which two whole numbers is $\frac{7}{4}$?

8. How much greater than a whole is …   a) $\frac{10}{7}$?   b) $\frac{6}{5}$?   c) $\frac{4}{3}$?   d) $\frac{11}{10}$?

NS6-64: Investigating Mixed & Improper Fractions

TEACHER:
Your students will need pattern blocks
for this exercise, or a copy of the patterns
block sheet from the Teacher's Guide.

 triangle rhombus 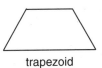 trapezoid

hexagon

NOTE: The blocks shown here are not actual size!

Euclid's bakery sells hexagonal pies. They sell pieces shaped like triangles, rhombi and trapezoids.

1. a) Shade $2\frac{5}{6}$ pies: b) How many pieces did you shade? _____

   c) Write an improper fraction from the amount of pie shaded: _____

2. Make a model of the pies below with pattern blocks. Place the smaller shapes on top of the
   hexagons, and then write a mixed and improper fractions for each pie.

   a)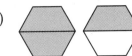

   Mixed Fraction: _____

   Improper Fraction: _____

   b)

   Mixed Fraction: _____

   Improper Fraction: _____

   c)

   Mixed Fraction: _____

   Improper Fraction: _____

3. Use the hexagon as the whole pie. Use the triangles, rhombuses and trapezoids as the pieces.
   Make a pattern block model of the fractions below. Then sketch your models on the grid.

   a) $2\frac{1}{2}$ 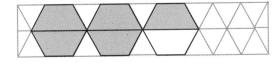   b) $3\frac{1}{2}$

   c) $2\frac{5}{6}$ 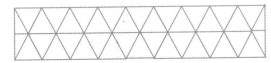   d) $2\frac{2}{3}$

4. Using the hexagon as the whole pie and the smaller pieces as the parts, make a pattern block model
   of the fractions. Sketch your model below.

   a)    b) $\frac{13}{6}$

   c) $\frac{7}{3}$ 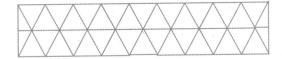   d) $\frac{14}{3}$

5. Using the trapezoid as the whole pie, and triangles as the pieces, make a pattern block model of the fractions. Sketch your models on the grid.

a) $\frac{5}{3}$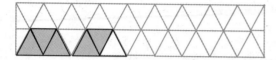

b) $\frac{7}{3}$

c) $1\frac{2}{3}$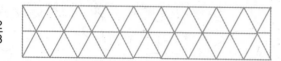

d) $2\frac{1}{3}$

Draw sketches (using the hexagon as the whole) to find the answers below.

6. Which fraction is greater: $2\frac{5}{6}$ or $\frac{15}{6}$?

7. Which fraction is greater: $3\frac{1}{3}$ or $\frac{11}{3}$?

8 Draw a picture to show $2 - \frac{1}{6}$.

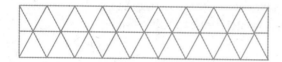

9. How much larger than a whole pie is $\frac{11}{6}$ of a pie?

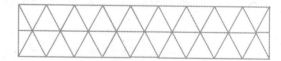

10. How much larger than two pies is $\frac{7}{3}$?

11. Ravi says $\frac{9}{6}$ pies is the same amount as $1\frac{1}{2}$ pies. Is he correct?

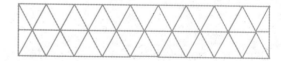

12. Jane sold $1\frac{2}{3}$ pies. Clara sold 11 pieces. (Each piece was $\frac{1}{6}$ of a pie). Who sold more pie?

13. Bernie ate $2\frac{2}{3}$ pizzas in June. How many third-sized pieces did he eat?

NS6-65: Equivalent Fractions

Aidan shades $\frac{2}{6}$ of the squares in an array:

He then draws heavy lines around the squares to group them into 3 equal groups:

He sees that $\frac{1}{3}$ of the squares are shaded.

The pictures show that two sixths are equal to one third: $\frac{2}{6} = \frac{1}{3}$

Two sixths and one third are **equivalent fractions**.

- -

1. Group the squares to make an equivalent fraction.

a)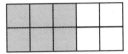

$\frac{6}{10} = \frac{}{5}$

b)

$\frac{4}{6} = \frac{}{3}$

c)

$\frac{10}{12} = \frac{}{6}$

2.  Write three equivalent fractions for the amount shaded here.

_____ _____ _____

3. Group the buttons to make an equivalent fraction.

a)

$\frac{4}{6} = \frac{}{}$

b)

$\frac{3}{6} = \frac{}{}$

c)

$\frac{2}{6} = \frac{}{}$

d)

$\frac{6}{9} = \frac{}{}$

e)

$\frac{6}{10} = \frac{}{}$

4. Cut each pie into smaller pieces to make an equivalent fraction.

a)

$\frac{2}{3} = \frac{}{6}$

b)

$\frac{2}{3} = \frac{}{9}$

c)

$\frac{3}{4} = \frac{}{12}$

5. a) Draw lines to cut the pies into …

4 pieces   6 pieces   8 pieces

b) Then fill in the numerators of the equivalent fractions.

$\frac{1}{2} = \frac{}{4} = \frac{}{6} = \frac{}{8}$

6. Draw shaded and unshaded circles (as in Question 3).
   Group the circles to show …

   a) six eighths is equivalent to three quarters

   b) four fifths is equivalent to eight tenths

Number Sense 2

Anne makes a model of $\frac{2}{5}$ using 15 squares as follows:

First she makes a model of $\frac{2}{5}$ using shaded and unshaded squares (leaving space between the squares).

Step 1: ▧ ▧ ☐ ☐ ☐

$\frac{2}{5}$ of the squares are shaded. She then adds squares one at a time until she has placed 15 squares.

Step 2: ▧▧ ▧▧ ☐☐ ☐☐ ☐☐

Step 3: ▧▧▧ ▧▧▧ ☐☐☐ ☐☐☐ ☐☐☐

From the picture, Anne can see that $\frac{2}{5}$ of a set of 15 squares is equivalent to $\frac{6}{15}$ of the set.

----

1. Draw a model of $\frac{2}{3}$ using 12 squares.
   The question is started for you.

   HINT: Place the extra squares beside the ones already drawn, one square at a time.

   ▧   ▧   ☐

2. Draw a model of $\frac{3}{5}$ using 10 squares.

   HINT: Start by making a model of $\frac{3}{5}$.

3. Draw a model of $\frac{3}{4}$ using 8 squares.

4. $\frac{5}{6}$ of a pizza is covered in olives **O**.
   $\frac{1}{3}$ of the pizza is covered in mushrooms ⇧.

   Each piece has a topping. Complete the picture.

   How many pieces are covered in olives **and** mushrooms? _____

5. If box C was cut into 12 pieces, how many pieces would you shade to make a fraction equivalent to A and B?

   A   B   C

NS6-67: Fractions of Whole Numbers

Dan has 6 cookies. He wants to give $\frac{2}{3}$ of his cookies to
his friends. To do so, he shares the cookies equally onto 3 plates:

There are 3 equal groups, so each group is $\frac{1}{3}$ of 6.

There are 2 cookies in each group, so $\frac{1}{3}$ of 6 is 2.

There are 4 cookies in two groups, so $\frac{2}{3}$ of 6 is 4.

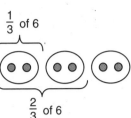

---

1. Write a fraction for the amount of dots shown. The first one has been done for you.

a)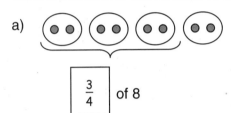

b)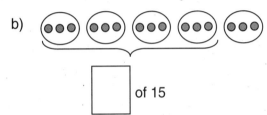

2. Fill in the missing numbers.

a)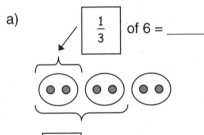

b)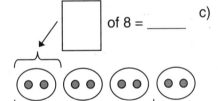

c)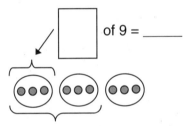

d)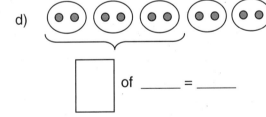

e)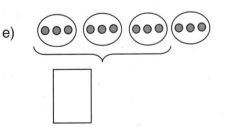

3. Draw a circle to show the given amount. The first one has been done for you.

a) $\frac{2}{3}$ of 6

b) $\frac{3}{4}$ of 8

c) $\frac{3}{5}$ of 10

d) $\frac{3}{4}$ of 12

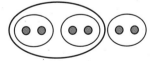

jump math
MULTIPLYING POTENTIAL

Number Sense 2

4. Fill in the correct number of dots in each circle, then draw a larger circle to show the given amount.

   a) $\frac{2}{3}$ of 12   ◯ ◯ ◯   b) $\frac{2}{3}$ of 9   ◯ ◯ ◯

5. Find the fraction of the whole amount by sharing the cookies equally.

   HINT: Draw the correct number of plates then place the cookies one at a time. Then circle the correct amount.

   a) Find $\frac{1}{4}$ of 8 cookies.   b) Find $\frac{1}{2}$ of 10 cookies.

   $\frac{1}{4}$ of 8 is _____ $\frac{1}{2}$ of 10 is _____

   c) Find $\frac{2}{3}$ of 6 cookies.   d) Find $\frac{3}{4}$ of 12 cookies.

   $\frac{2}{3}$ of 6 is _____ $\frac{3}{4}$ of 12 is _____

6. Andy finds $\frac{2}{3}$ of 12 as follows:

   Step 1   He finds $\frac{1}{3}$ of 12 by dividing 12 by 3:

   12 ÷ 3 = 4 (4 is $\frac{1}{3}$ of 12)

   Step 2   Then he multiplies the result by 2: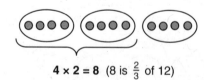

   4 × 2 = 8 (8 is $\frac{2}{3}$ of 12)

   Find the following amounts using Andy's method.

   a) $\frac{2}{3}$ of 9   b) $\frac{3}{4}$ of 8   c) $\frac{2}{3}$ of 15   d) $\frac{2}{5}$ of 10

   _____   _____   _____   _____

   e) $\frac{3}{5}$ of 25   f) $\frac{2}{7}$ of 14   g) $\frac{1}{6}$ of 18   h) $\frac{1}{2}$ of 12

   _____   _____   _____   _____

   i) $\frac{3}{4}$ of 12   j) $\frac{2}{3}$ of 21   k) $\frac{3}{8}$ of 16   l) $\frac{3}{7}$ of 21

   _____   _____   _____   _____

Number Sense 2

NS6-67: Fractions of Whole Numbers (continued)

7. a) Shade $\frac{2}{5}$ of the boxes.   b) Shade $\frac{2}{3}$ of the boxes.   c) Shade $\frac{3}{4}$ of the boxes.

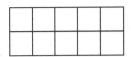   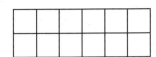   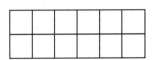

8. a) Shade $\frac{1}{4}$ of the boxes. Draw stripes in $\frac{1}{6}$ of the boxes.   b) Shade $\frac{1}{3}$ of the boxes. Draw stripes in $\frac{1}{6}$ of the boxes. Put dots in $\frac{1}{8}$ of the boxes.

   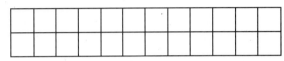

9. 15 children are on a bus. $\frac{3}{5}$ are girls. How many girls are on the bus? _____

10. A kilogram of lichees costs $8. How much would $\frac{3}{4}$ of a kilogram cost? _____

11. Gerald has 12 apples. He gives away $\frac{3}{4}$ of the apples.

   a) How many apples did he give away? _____   b) How many did he keep? _____

12.   Shade $\frac{1}{3}$ of the squares.

   Draw stripes in $\frac{1}{6}$ of the squares.

   How many squares are blank?

13. Evelyn has 20 marbles.
   $\frac{2}{5}$ are blue. $\frac{1}{4}$ are yellow. The rest are green.
   How many are green?

14. Ed started studying at 9:10. He studied for $\frac{2}{3}$ of an hour.
   At what time did he stop studying?

15. Marion had 36 stickers.
   She kept $\frac{1}{6}$ for herself and divided the rest evenly among 5 friends.
   How many stickers did each friend get?

16. Which is longer: 17 months or $1\frac{3}{4}$ years?

17. Linda had 12 apples.
   She gave $\frac{1}{4}$ of them to Nandita. She gave 2 to Amy.
   She says that she has half left. Is she correct?

NS6-68: Reducing Fractions

A fraction is reduced to **lowest terms** when the only whole number that will divide into its numerator and denominator is the number 1. $\frac{2}{4}$ is not in lowest terms (because 2 divides into 2 and 4) but $\frac{1}{2}$ is in lowest terms.

You can reduce a fraction to lowest terms by dividing a set of counters representing the fraction into equal groups. (You may have to group counters several times before the fraction is in lowest terms.)

Step 1: Count the number of counters in each group.

Step 2: Divide the numerator and denominator of the fraction by the number of counters in each group.

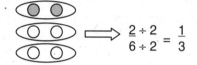 $\frac{2 \div 2}{6 \div 2} = \frac{1}{3}$ 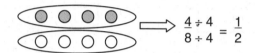 $\frac{4 \div 4}{8 \div 4} = \frac{1}{2}$

----

1. Reduce these fractions by grouping.

   a) 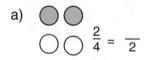 $\frac{2}{4} = \frac{}{2}$

   b) $\frac{3}{9} = \frac{}{3}$

   c) $\frac{4}{6} = \frac{}{3}$

2. Show how you would reduce the fractions by dividing.

   a) $\frac{2 \div}{4 \div}$ = ———

   b) $\frac{3 \div}{9 \div}$ = ———

   c) $\frac{4 \div}{6 \div}$ = ———

3. Reduce the fractions below by dividing.

   a) $\frac{2}{10}$ = —

   b) $\frac{2}{6}$ = —

   c) $\frac{2}{8}$ = —

   d) $\frac{2}{12}$ = —

   e) $\frac{3}{9}$ = —

   f) $\frac{3}{15}$ = —

   g) $\frac{4}{12}$ = —

   h) $\frac{6}{9}$ = —

   i) $\frac{4}{6}$ = —

   j) $\frac{10}{15}$ = —

   k) $\frac{20}{25}$ = —

   l) $\frac{8}{12}$ = —

4. The pie chart show how the children in a Grade 6 class get to school.
   What fraction of the children …

   a) walk to school? ⬚

   b) take the bus? ⬚

   c) skateboard? ⬚

   d) cycle? ⬚

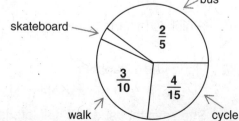

NS6-69: Lowest Common Multiples in Fractions

1.

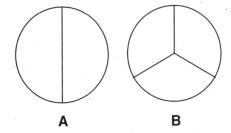

A   B

How many pieces are in pie A? _____

How many pieces are in pie B? _____

Find the LCM of the number of
pieces in pies A and B: **LCM =** _____

Cut pie A and pie B into this many pieces.

How many pieces did you cut each piece
of pie A into? _____

How many pieces did you cut each piece
of pie B into? _____

2.

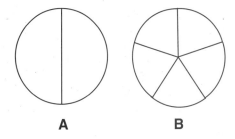

A   B

How many pieces are in pie A? _____

How many pieces are in pie B? _____

Find the LCM of the number of
pieces in pies A and B: **LCM =** _____

Cut pie A and pie B into this many pieces.

How many pieces did you cut each piece
of pie A into? _____

How many pieces did you cut each piece
of pie B into? _____

3.

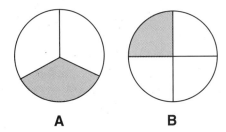

A   B

Write the name of each fraction.

A - _____   B - _____

Find the LCM of the number of
pieces in pies A and B: **LCM =** _____

Cut each pie into the same number of pieces,
given by the LCM.

Now write a new name for each fraction.

A - _____   B - _____

4.

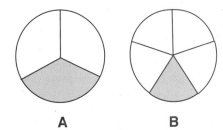

A   B

Write the name of each fraction.

A - _____   B - _____

Find the LCM of the number of
pieces in pies A and B: **LCM =** _____

Cut each pie into the same number of pieces,
given by the LCM.

Now write a new name for each fraction.

A - _____   B - _____

jump math
MULTIPLYING POTENTIAL.

Number Sense 2

5.

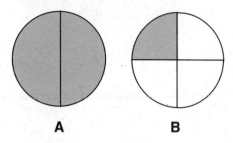

A   **B**

Write the name of each fraction.

A - _____   B - _____

Find the LCM of the number of pieces in pies A and B: **LCM** = _____

Cut each pie into the same number of pieces, given by the **LCM**.

Now write a new name for each fraction.

A - _____   B - _____

6.

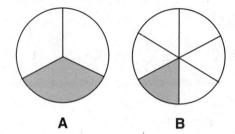

A   **B**

Write the name of each fraction.

A - _____   B - _____

Find the LCM of the number of pieces in pies A and B: **LCM** = _____

Cut each pie into the same number of pieces, given by the **LCM**.

Now write a name for each fraction.

A - _____   B - _____

7.

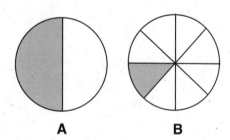

A   **B**

Write the name of each fraction.

A - _____   B - _____

Find the LCM of the number of pieces in pies A and B: **LCM** = _____

Cut each pie into the same number of pieces, given by the LCM.

Now write a new name for each fraction.

A - _____   B - _____

8.

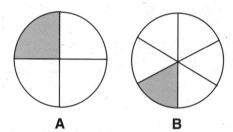

A   **B**

Write the name of each fraction.

A - _____   B - _____

Find the LCM of the number of pieces in pies A and B: **LCM** = _____

Cut each pie into the same number of pieces, given by the LCM.

Now write a new name for each fraction.

A - _____   B - _____

NS6-70: Comparing and Ordering Fractions

Use the fraction strips below to answer Questions 1 to 3.

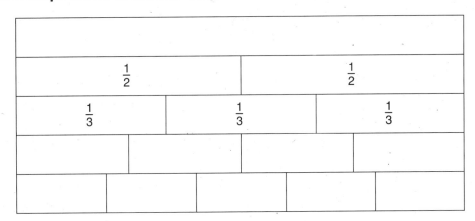

1. Fill in the missing numbers on the fraction strips above. Then write > (greater than) or < (less than) between each pair of numbers below.

   a) $\frac{2}{3} \square \frac{3}{5}$ b) $\frac{3}{4} \square \frac{1}{2}$ c) $\frac{2}{5} \square \frac{3}{4}$ d) $\frac{3}{4} \square \frac{4}{5}$

2. Circle the fractions that are greater than $\frac{1}{3}$.

   $\frac{3}{5}$ $\frac{2}{5}$ $\frac{1}{2}$

3. Circle the fractions that are greater than $\frac{1}{2}$.

   $\frac{2}{3}$ $\frac{2}{5}$ $\frac{3}{4}$

4. Turn each fraction into an equivalent fraction so that both fractions have the same denominator (by finding the LCM of both denominators). Then write =, <, or > between the two fractions.

   a) $\frac{5 \times 1}{5 \times 2} \square \frac{7}{10}$ b) $\frac{1}{2} \square \frac{3}{10}$ c) $\frac{3}{4} \square \frac{7}{8}$ d) $\frac{11}{20} \square \frac{2}{5}$

   $\frac{5}{10} \boxed{<} \frac{7}{10}$ $\square$ $\square$ $\square$

   e) $\frac{2}{3} \square \frac{4}{5}$ f) $\frac{1}{2} \square \frac{2}{3}$ g) $\frac{3}{4} \square \frac{2}{3}$ h) $\frac{2}{3} \square \frac{6}{9}$

   $\square$ $\square$ $\square$ $\square$

   i) $\frac{3}{4} \square \frac{4}{5}$ j) $\frac{1}{7} \square \frac{5}{21}$ k) $\frac{17}{35} \square \frac{3}{5}$ l) $\frac{4}{5} \square \frac{5}{6}$

   $\square$ $\square$ $\square$ $\square$

Number Sense 2

1. Write the fractions in order from least to greatest.

   HINT: First write each fraction with the same denominator.

   a) $\frac{1}{2}$ $\frac{2}{5}$ $\frac{3}{10}$

   b) $\frac{1}{3}$ $\frac{5}{6}$ $\frac{1}{2}$

   c) $\frac{5}{8}$ $\frac{1}{2}$ $\frac{3}{4}$

   a) $\boxed{\frac{}{10}}$ $\boxed{\frac{}{10}}$ $\boxed{\frac{3}{10}}$

   _____   _____   _____

2. Write the fractions in the boxes in order from least to greatest.

   $\boxed{0}$ $\boxed{}$ $\boxed{}$ $\boxed{}$ $\boxed{}$ $\boxed{}$ $\boxed{}$ $\boxed{}$ $\boxed{}$ $\boxed{1}$

   $\frac{1}{10}$ $\frac{2}{5}$ $\frac{9}{10}$ $\frac{4}{5}$ $\frac{3}{5}$ $\frac{3}{10}$ $\frac{1}{5}$ $\frac{1}{2}$ $\frac{7}{10}$

3. Equivalent fractions are said to be in the same **family**. Write two fractions in the same family as the fraction in each triangle.

   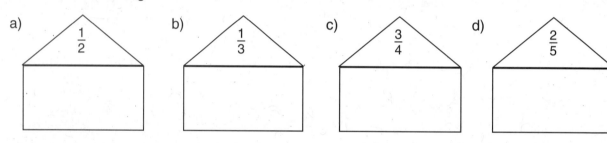

   a) $\frac{1}{2}$ b) $\frac{1}{3}$ c) $\frac{3}{4}$ d) $\frac{2}{5}$

4. In each question, circle the <u>pair</u> of fractions that are in the same family.

   a) $\frac{1}{2}$ $\frac{4}{6}$ $\frac{5}{10}$

   b) $\frac{2}{3}$ $\frac{4}{6}$ $\frac{1}{4}$

   c) $\frac{3}{15}$ $\frac{16}{20}$ $\frac{4}{5}$

5. Fill in the missing fractions in the sequence.
   HINT: Give all the fractions the same denominator.

   $\frac{1}{6}$ $\frac{1}{3}$ $\frac{1}{2}$ $\boxed{}$ $\frac{5}{6}$ $\boxed{}$

6. Explain how you know $\frac{1}{3}$ is greater than $\frac{1}{8}$.

7. Find 2 fractions from the fraction family of $\frac{4}{12}$ with numerators smaller than 4.

8. Find 5 fractions from the fraction family of $\frac{12}{24}$ with numerators smaller than 12.

9. A recipe for soup calls for $\frac{2}{3}$ of a can of tomatoes.

   A recipe for spaghetti sauce calls for $\frac{5}{6}$ of a can.
   Which recipe uses more tomatoes?

NS6-72: Word Problems

Answer the following questions in your notebook.

1. Anne had 1 hour for lunch.

   She played for $\frac{3}{5}$ of an hour and read for $\frac{1}{10}$ of an hour.

   a) How many minutes did she have left to eat lunch?

   b) What fraction of an hour was this?

Colour	Number of Walls Painted
White	10
Yellow	5
Blue	4
Green	1

2. a) What fraction of the walls were painted green?

   b) What colour was used to paint one fifth of the walls?

   c) What colour was used to paint one half of the walls?

3. Charles left for school at 7:10 am. He walked for $\frac{2}{5}$ of an hour to his friend's house, and then another $\frac{3}{5}$ of an hour to school. At what time did he arrive at school?

4. The pie chart shows the times of day when a lizard is active.

    Awake but Inactive

    Asleep

    Awake and Active

   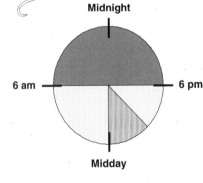

   a) What fraction of the day is the lizard…

      i) awake but inactive   ii) asleep   iii) awake and active

   b) How many hours a day is the lizard…

      i) awake but inactive   ii) asleep   iii) awake and active

5. John, Brian, Eldad, and Ahmed bought some pizzas.
   They ate the following fractions
   of a pizza (in no particular order): $\frac{2}{10}$, $\frac{2}{5}$, $\frac{7}{10}$, $\frac{6}{5}$
   There were 5 slices left over.

   a) How many pizzas did they buy?

   b) • Eldad ate $\frac{1}{5}$ of a pizza.

       • Ahmed ate $\frac{7}{10}$, which was 3 more slices than Brian.

       • John ate more than one pizza.

       How much pizza did each boy eat?

6. A box holds 9 cans. The box is $\frac{2}{3}$ full.
   How many cans are in the box?

Number Sense 2

NS6-73: Decimal Hundredths

Fractions with denominators that are multiples of ten (tenths, hundredths) commonly appear in units of measurement.

- A millimetre is a tenth of a centimetre (10 mm = 1 cm)
- A centimetre is a tenth of a decimetre (10 cm = 1 dm)
- A decimetre is a tenth of a metre (10 dm = 1 m)
- A centimetre is a hundredth of a metre (100 cm = 1 m)

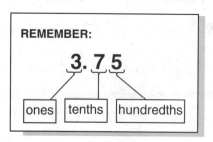

REMEMBER:

3.75

ones | tenths | hundredths

Decimals are short forms for fractions: **.73** has 7 tenths (= 70 hundredths) plus 3 more hundredths.

----

1. Write a fraction and a decimal for each picture.

a)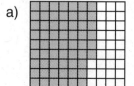

b)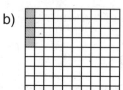

c)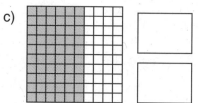

2. Convert the fraction to a decimal. Then shade.

a) $\dfrac{39}{100} =$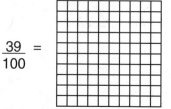

b) $\dfrac{65}{100} =$

c) $\dfrac{7}{100} =$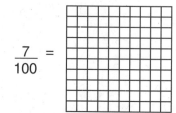

3. The picture shows a floor plan of a zoo. Write a fraction and a decimal for each shaded part:

Key:

▨ reptiles

▩ birds

⸬ amphibians

☐ paths

4. Make your own floor plan for a zoo. Write a fraction and a decimal for each shaded part:

NS6-74: Tenths and Hundredths

1. Draw lines around the columns to show tenths as in a). Then, write a fraction and a decimal to represent the number of shaded squares.

a) 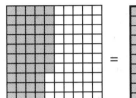 =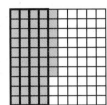

47 hundredths = 4 tenths ___ hundredths

$$\frac{47}{100} = .\;\underline{\;4\;}\;\underline{\;7\;}$$

b) 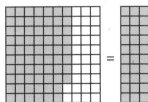 =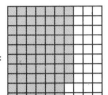

___ hundredths = ___ tenths ___ hundredths

$$\overline{100} = .\;\underline{\;\;\;}\;\underline{\;\;\;}$$

c) 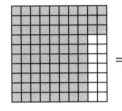 =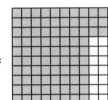

___ hundredths = ___ tenths ___ hundredths

$$\overline{100} = .\;\underline{\;\;\;}\;\underline{\;\;\;}$$

d)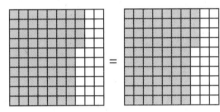

___ hundredths = ___ tenths ___ hundredths

$$\overline{100} = .\;\underline{\;\;\;}\;\underline{\;\;\;}$$

2. Fill in the blanks.

a) 43 hundredths = ___ tenths ___ hundredths

$$\frac{43}{100} = .\;\underline{\;4\;}\;\underline{\;3\;}$$

b) 28 hundredths = ___ tenths ___ hundredths

$$\overline{100} = .\;\underline{\;\;\;}\;\underline{\;\;\;}$$

c) 66 hundredths = ___ tenths ___ hundredths

$$\overline{100} = .\;\underline{\;\;\;}\;\underline{\;\;\;}$$

d) 84 hundredths = ___ tenths ___ hundredths

$$\overline{100} = .\;\underline{\;\;\;}\;\underline{\;\;\;}$$

e) 9 hundredths = ___ tenths ___ hundredths

$$\overline{100} = .\;\underline{\;\;\;}\;\underline{\;\;\;}$$

f) 30 hundredths = ___ tenths ___ hundredths

$$\overline{100} = .\;\underline{\;\;\;}\;\underline{\;\;\;}$$

3. Describe each decimal in two ways.

a) .52 = __5__ tenths __2__ hundredths

= _____52 hundredths_____

b) .55 = ___ tenths ___ hundredths

= _____

c) .40 = ___ tenths ___ hundredths

= _____

d) .23 = ___ tenths ___ hundredths

= _____

e) .02 = ___ tenths ___ hundredths

= _____

f) .18 = ___ tenths ___ hundredths

= _____

1. Fill in the chart below. The first one has been done for you.

Drawing	Fraction	Decimal	Equivalent Decimal	Equivalent Fraction	Drawing
	$\frac{4}{10}$	0.4	0.40	$\frac{40}{100}$	

2. Write a fraction for the number of <u>hundredths</u>. Then count the shaded columns and write a fraction for the number of <u>tenths</u>.

a)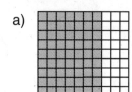

$\overline{100} = \overline{10}$

b)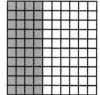

$\overline{100} = \overline{10}$

c)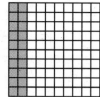

$\overline{100} = \overline{10}$

d)

$\overline{100} = \overline{10}$

3. Fill in the missing numbers. **REMEMBER:** $\frac{10}{100} = \frac{1}{10}$

a) $.5 = \frac{5}{10} = \frac{}{100} = .\,\underline{\ }\,\underline{\ }$

b) $.\,\underline{\ } = \frac{3}{10} = \frac{}{100} = .30$

c) $.\,\underline{\ } = \frac{9}{10} = \frac{}{100} = .90$

d) $.\,\underline{\ } = \frac{8}{10} = \frac{}{100} = .\,\underline{\ }\,\underline{\ }$

e) $.\,\underline{\ } = \frac{}{10} = \frac{40}{100} = .\,\underline{\ }\,\underline{\ }$

f) $.\,\underline{\ } = \frac{}{10} = \frac{70}{100} = .\,\underline{\ }\,\underline{\ }$

g) $.\,\underline{\ } = \frac{4}{10} = \frac{}{100} = .\,\underline{\ }\,\underline{\ }$

h) $.\,\underline{\ } = \frac{6}{10} = \frac{}{100} = .\,\underline{\ }\,\underline{\ }$

i) $.3 = \frac{}{10} = \frac{}{100} = .\,\underline{\ }\,\underline{\ }$

NS6-76: Decimals, Money and Measurements

A **dime** is **one tenth** of a dollar. A **penny** is one **hundredth** of a dollar.

1. Express the value of each decimal in four different ways.

a) .64

    6 dimes 4 pennies

    6 tenths 4 hundredths

    64 pennies

    64 hundredths

b) .72

c) . 43

d) .04

2. A **decimeter** is **one tenth** of a metre. A **centimetre** is a **hundredth** of a metre. Express the value of each measurement in four different ways.

a) .28 m

    2 decimetres 8 centimetres

b) .13 m

3. Express the value of each decimal in 4 different ways.
   HINT: First add a zero in the hundredths place.

a) .6 _____ dimes _____ pennies

    _____ tenths _____ hundredths

    _____ pennies

    _____ hundredths

b) .8 _____ dimes _____ pennies

    _____ tenths _____ hundredths

    _____ pennies

    _____ hundredths

4. Express the value of each decimal in four different ways. Then circle the greater number.

a) .27 _____ dimes _____ pennies

    _____ tenths _____ hundredths

    _____ pennies

    _____ hundredths

b) .3 _____ dimes _____ pennies

    _____ tenths _____ hundredths

    _____ pennies

    _____ hundredth

5. George says .63 is greater than .8 because 63 is greater than 8. Can you explain his mistake?

jump math
MULTIPLYING POTENTIAL

Number Sense 2

1. Fill in the missing numbers.

a)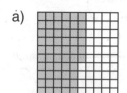
b)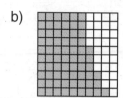
c)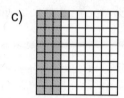
d)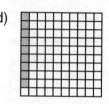

tenths	hundredths

tenths	hundredths

tenths	hundredths

tenths	hundredths

$\overline{100}$ = . ___ ___
        tenths hundredths

$\overline{100}$ = . ___ ___

$\overline{100}$ = . ___ ___

$\overline{100}$ = . ___ ___

2. Write the following decimals as fractions.

a) .3 = $\overline{10}$ b) .5 = $\overline{10}$ c) .8 = $\overline{10}$ d) .2 = $\overline{10}$ e) .1 = $\overline{10}$

f) .34 = $\overline{100}$ g) .39 = $\overline{100}$ h) .77 = $\overline{100}$ i) .86 = $\overline{100}$ j) .61 = $\overline{100}$

k) .7 =   l) .34 =   m) .06 =   n) .4 =   o) .04 =

p) .6 =   q) .46 =   r) .25 =   s) .93 =   t) .06 =

3. Change the following fractions to decimals.

a) $\frac{2}{10}$ = . ___   b) $\frac{4}{10}$ = . ___   c) $\frac{3}{10}$ = . ___   d) $\frac{9}{10}$ = . ___

e) $\frac{93}{100}$ = . ___ ___   f) $\frac{78}{100}$ = . ___ ___   g) $\frac{66}{100}$ = . ___ ___   h) $\frac{5}{100}$ = . ___ ___

4. Circle the equalities that are incorrect.

a) .36 = $\frac{36}{100}$ b) .9 = $\frac{9}{100}$ c) .6 = $\frac{6}{10}$ d) $\frac{27}{100}$ = .27   e) $\frac{3}{100}$ = .03

f) .75 = $\frac{74}{100}$ g) .40 = $\frac{40}{10}$ h) .75 = $\frac{75}{100}$ i) .08 = $\frac{8}{100}$ j) .03 = $\frac{3}{10}$

5. Write as a decimal.

a) 8 tenths 2 hundredths =   b) 0 tenths 9 hundredths =

6. Write .46 as a fraction in lowest terms. Explain how you found your answer.

A hundreds block may be used to represent a whole. 10 is one tenth of 100, so a tens block represents one tenth of the whole. 1 is one hundredth of 100, so a ones block represents one hundredth of the whole.

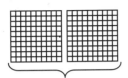

2 wholes   3 tenths   4 hundredths

ones   hundredths

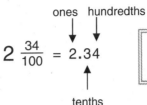

$2\frac{34}{100} = 2.34$

tenths

NOTE: A mixed fraction can be written as a decimal.

---

1. Write a mixed fraction and a decimal for the base ten models below.

a)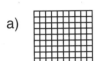

b)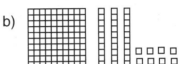

c)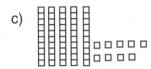

d)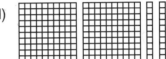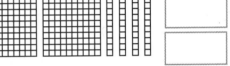

e)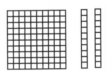

2. Draw a base ten model for the following decimals.

a) 3.21

b) 1.62

3. Write a decimal and a mixed fraction for each of the pictures below.

a)

b)

4. Write a decimal for each of the mixed fractions below.

a) $1\frac{32}{100} =$

b) $2\frac{71}{100} =$

c) $8\frac{7}{10} =$

d) $4\frac{27}{100} =$

e) $3\frac{7}{100} =$

f) $17\frac{8}{10} =$

g) $27\frac{1}{10} =$

h) $38\frac{5}{100} =$

5. Which decimal represents a greater number?
Explain your answer with a picture.

a) 6 tenths or 6 hundredths?

b) .8 or .08?

c) 1.02 or 1.20?

This number line is divided into tenths. The number represented by Point A is $2\frac{3}{10}$ or 2.3.

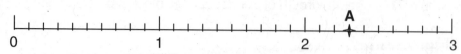

1. Write a fraction or a mixed fraction for each point.

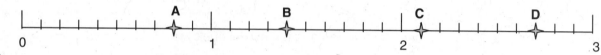

A: _____ **B:** _____ **C:** _____ **D:** _____

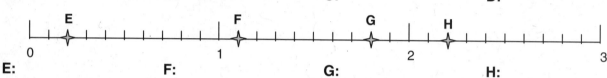

E: _____ **F:** _____ **G:** _____ **H:** _____

2. Mark each point with an 'X' and label the point with the correct letter.

A. 1.3   **B.** 2.7   **C.** .70   **D.** 1.1

                                                              E. $2\frac{1}{10}$

F. one and three tenths   **G.** nine tenths   **H.** one and one tenth   **I.** two decimal nine

3. Write the name of each point as a decimal in words.

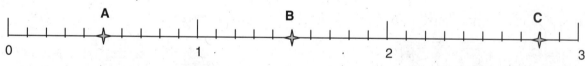

A. _____   **B.** _____   **C.** _____

4. Mark the decimals on the number lines.

a) **0.6**   b) **1.2**

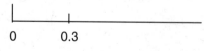   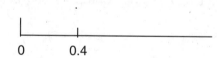

BONUS
5. Mark the following fractions and decimals on the number line.

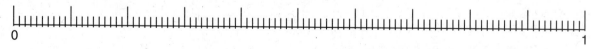

A. .72   **B.** $\frac{34}{100}$ **C.** .05   **D.** $\frac{51}{100}$

Number Sense 2

NS6-80: Comparing and Ordering Fractions and Decimals

1.

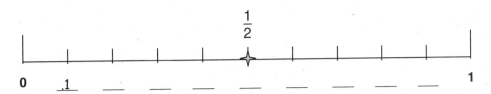

a) Write a decimal for each point marked on the number line. (The first decimal is written for you.)

b) Which decimal is equal to one half? $\frac{1}{2}$ = _____

2. Use the number line in Question 1 to say whether each decimal is closer to "zero", "a half" or "one".

a) .3 is closer to _____

b) .7 is closer to _____

c) .8 is closer to _____

d) .9 is closer to _____

e) .1 is closer to _____

f) .2 is closer to _____

3.

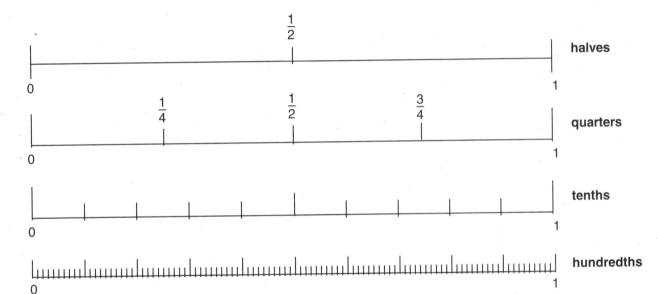

Use the number lines above to compare the numbers given. Write < (less than) or > (greater than) between each pair of numbers.

a) 0.7 ☐ $\frac{3}{4}$

b) 0.4 ☐ $\frac{7}{10}$

c) 0.8 ☐ $\frac{1}{2}$

d) 0.2 ☐ $\frac{1}{4}$

e) 0.4 ☐ $\frac{1}{2}$

f) 0.35 ☐ $\frac{1}{4}$

g) 0.07 ☐ $\frac{1}{2}$

h) $\frac{3}{4}$ ☐ .65

4. Which whole number is each decimal or mixed fraction closest to: "zero", "one", "two," or "three"?

a) 2.4 is closest to _____

b) 2.8 is closest to _____

c) $1\frac{3}{10}$ is closest to _____

jump math
MULTIPLYING POTENTIAL

Number Sense 2

1. Write the numbers in order by first changing each decimal to a fraction with a denominator of 10.
   NOTE: Show your work beside each number.

   a) 0.6 $\frac{6}{10}$ 0.7   0.4   b) 1.2 $1\frac{2}{10}$ 3.7   3.5   c) 4.7   4.5 $4\frac{3}{10}$

   _____   _____   _____

2. Ali says: "To compare .6 and .42, I add a zero to .6:

   .6 = 6 tenths = 60 hundredths = .60

   60 (hundredths) is greater than 42 (hundredths).

   So .6 is greater than .42."

   Add a zero to the decimal expressed in tenths. Then circle the greater number in each pair.

   a) .4   .32   b) .72   .8   c) .32   .2

3. Write each decimal as a fraction with denominator 100 by first adding a zero to the decimal.

   a) .7 = [.70] = [$\frac{70}{100}$]   b) .9 = [ ] = [ ]   c) .1 = [ ] = [ ]

4. Write the numbers in order from least to greatest by first changing all of the decimals to fractions with denominator 100.

   a) .3   .9   .45   b) $\frac{37}{100}$ .8   .32   c) 1.4 $1\frac{34}{100}$ 1.35

   [$\frac{30}{100}$] [ ] [ ]   [ ] [ ] [ ]   [ ] [ ] [ ]

   _____   _____   _____

5. Change $\frac{27}{10}$ to a mixed fraction by shading the correct number of pieces.

            Mixed Fraction: _____

6. Change the following improper fractions to mixed fractions.

   a) $\frac{25}{10}$ b) $\frac{37}{10}$ c) $\frac{86}{10}$ d) $\frac{60}{10}$ e) $\frac{186}{100}$ f) $\frac{175}{100}$

7. Change the following improper fractions to decimals by first writing them as mixed fractions.

   a) $\frac{35}{10} = 3\frac{5}{10} = 3.5$ b) $\frac{38}{10}$ c) $\frac{87}{10}$ d) $\frac{53}{10}$ e) $\frac{153}{100}$ f) $\frac{342}{100}$

8. Which is greater, $\frac{23}{10}$ or 2.4? Explain. | 9. Write 5 decimals greater than 1.32 and less than 1.4

10. 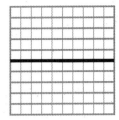 Shade $\frac{1}{2}$ of the squares. Write 2 fractions and 2 decimals for $\frac{1}{2}$.

Fractions: $\frac{1}{2}$ = $\frac{}{10}$ = $\frac{}{100}$

Decimals: $\frac{1}{2}$ = .____ = .____

11. 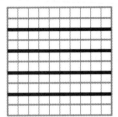 Shade $\frac{1}{5}$ of the boxes. Write 2 fractions and 2 decimals for $\frac{1}{5}$.

Fractions: $\frac{1}{5}$ = $\frac{}{10}$ = $\frac{}{100}$

Decimals: $\frac{1}{5}$ = .____ = .____

12. Write equivalent fractions.

a) $\frac{2}{5}$ = $\frac{}{10}$ = $\frac{}{100}$ b) $\frac{3}{5}$ = $\frac{}{10}$ = $\frac{}{100}$ c) $\frac{4}{5}$ = $\frac{}{10}$ = $\frac{}{100}$

13. 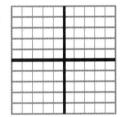 Shade $\frac{1}{4}$ of the squares. Write a fraction and a decimal for $\frac{1}{4}$ and $\frac{3}{4}$.

Fraction: $\frac{1}{4}$ = $\frac{}{100}$ Decimal: $\frac{1}{4}$ = .____

Fraction: $\frac{3}{4}$ = $\frac{}{100}$ Decimal: $\frac{3}{4}$ = .____

14. Circle the greater number.
HINT: First change all fractions and decimals to fractions with denominator 100.

a) $\frac{3}{4}$. 72   b) $\frac{1}{2}$ .53   c) $\frac{3}{5}$. 87

  ◯   ◯ ◯   ◯ ◯

15. Write the numbers in order from least to greatest. Explain how you found your answer.

a) .8   .42 $\frac{3}{4}$ b) $\frac{1}{2}$ $\frac{4}{5}$ .35   c) $\frac{3}{5}$ .45 $\frac{1}{2}$

16. How does knowing that $\frac{1}{4}$ = 0.25 help you find the decimal form of $\frac{3}{4}$?

17. Explain how you know 0.65 is greater that $\frac{1}{2}$.

NS6-82: Thousandths

If a thousands cube is used to represent a whole number, then a hundreds block represents a tenth, a tens block represents a hundredth, and a ones block represents a thousandth of a whole.

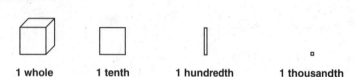

1 whole   1 tenth   1 hundredth   1 thousandth

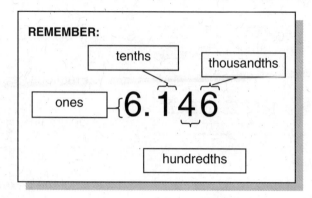

REMEMBER:

1. Beside each number, write the place value of the underlined digit.

   a)   3.2<u>7</u>4

   b)   9.27<u>3</u>

   c)   2.<u>5</u>37

   d)   7.12<u>9</u>

   e)   <u>5</u>.214

   f)   8.9<u>7</u>8

2. Write the following numbers into the place value chart. The first one has been done for you.

	ones	tenths	hundredths	thousandths
a) 6.512	6	5	1	2
c) 7.03				
e) 1.763				
g) 6.38				
i) 5.813				

	ones	tenths	hundredths	thousandths
b) 6.354				
d) 1.305				
f) 0.536				
h) 8				
j) 0.13				

3. Write the following decimals as fractions.

   a) .652 =

   b) .372 =

   c) .20 =

   d) .002 =

4. Write each decimal in expanded form.

   a) .237 = __2 tenths + 3 hundredths + 7 thousandths__

   b) .325 = _____

   c) 6.336 = _____

5. Write the following fractions as decimals.

   a) $\dfrac{49}{100}$ =

   b) $\dfrac{50}{100}$ =

   c) $\dfrac{758}{1000}$ =

   d) $\dfrac{25}{1000}$ =

6. Compare each pair of decimals by writing < or > in the box.
   HINT: Add zeroes wherever necessary to give each number the same number of digits after the decimal point.

   a) .375 ☐ .378

   b) .233 ☐ .47

   c) .956 ☐ .1

   d) .27 ☐ .207

   e) .7 ☐ .32

   f) .8 ☐ .516

Number Sense 2

NS6-83: Adding Hundredths

1. Write a fraction for each shaded part. Then add the fractions, and shade your answer. The first one has been done for you.

a) 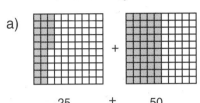 + =

$$\frac{25}{100} \quad + \quad \frac{50}{100} \quad = \quad \frac{75}{100}$$

b) + =

c) 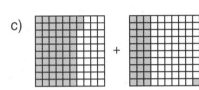 + =

d) 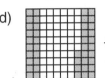 + =

2. Write the decimals that correspond to the fractions in Question 1 above.

a) .25 + .50 = .75	b)
c)	d)

3. Add the decimals by lining up the digits. Be sure that your final answer is expressed as a decimal.

a) 0.32 + 0.57 =   b) 0.91 + 0.04 =   c) 0.42 + 0.72 =   d) 0.22 + 0.57 =

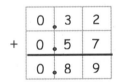 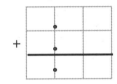 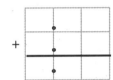 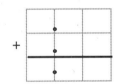

e) 0.3 + 0.36 =   f) 0.5 + 0.48 =   g) 0.81 + 0.58 =   h) 0.46 + 0.22 =

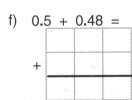 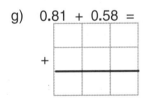 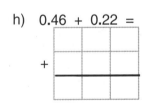

4. Line up the decimals and add the following numbers.

a) 4.32 + 2.17   b) 3.64 + 5.23   c) 9.46 + 3.12   d) 0.87 + 0.02   e) 4.8 + 0.31

5. Each wing of a butterfly is 3.72 cm wide. It's body is .46 cm wide. How wide is the butterfly?

6. Anne made punch by mixing .63 litres of juice with .36 litres of ginger ale. How many litres of punch did she make?

jump math
MULTIPLYING POTENTIAL.

NS6-84: Subtracting Hundredths

1. Subtract by crossing out the correct number of boxes.

a)
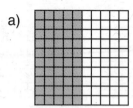

$$\frac{50}{100} - \frac{30}{100} =$$

b)
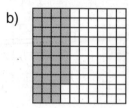

$$\frac{38}{100} - \frac{12}{100} =$$

c)
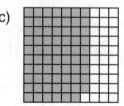

$$\frac{69}{100} - \frac{34}{100} =$$

2. Write the decimals that correspond to the fractions in Question 1 above.

a) .50 - .30 = .20	b)	c)

3. Subtract the decimals by lining up the digits.

a) 0.53 − 0.21 =

	0	5	3
−	0	2	1
	0	3	2

b) 0.88 − 0.34 =
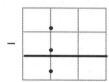

c) 0.46 − 0.23 =
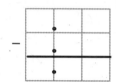

d) 0.75 − 0.21 =
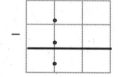

e) 0.33 − .17 =
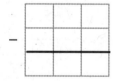

f) 0.64 − 0.38 =
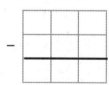

g) 0.92 − 0.59 =

h) 0.53 − 0.26 =
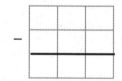

i) 1.00 − .82 =
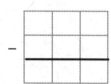

j) 1.00 − 0.36 =
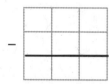

k) 1.00 − 0.44 =
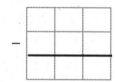

l) 1.00 − 0.29 =
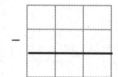

4. Subtract the following decimals.

   a) .82 − .45   b) .97 − .38   c) .72 − .64   d) .31 − .17

   e) .58 − .3   f) .62 − .6   g) .98 − .03   h) .53 − .09

5. Find the missing decimal in each of the following.

   a) 1 = .35 + ☐   b) 1 = .72 + ☐   c) 1 = .41 + ☐

jump math
MULTIPLYING POTENTIAL

Number Sense 2

NS6-85: Adding and Subtracting Decimals

1. Add by drawing a base ten model. Then, using the chart provided, line up the decimal points and add.
   NOTE: Use a hundreds block for a whole and a tens block for one tenth.

   a) 1.23 + 1.12

   b) 1.46 + 1.33

ones	tenths	hundredths
+		

ones	tenths	hundredths
+		

2. Draw a model of the greater number. Then subtract by crossing out blocks as shown in part a).

   a) 2.35 − 1.12 = 1.23

   b) 3.24 − 2.11

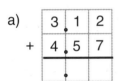 = 1.23

3. Add or subtract.

   a)
   ```
     3 . 1 2
   + 4 . 5 7
   ```

   b)
   ```
     5 . 8 9
   + 1 . 3 4
   ```

   c)
   ```
     3 . 8 6
   − 2 . 1 5
   ```

   d)
   ```
     4 . 2 3
   − 2 . 1 9
   ```

   e)
   ```
     1 8 . 0 5
   − 1 2 . 7 3
   ```

4. Subtract each pair of numbers by lining up the decimal points.

   a) 7.87 − 4.03   b) 9.74 − 6.35   c) 2.75 − .28   d) 28.71 − 1.4   e) 17.9 − 4.29

5. 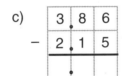 The average temperature in Jakarta is 30.33°C and, in Toronto, it is 11.9°C.
   How much warmer is Jakarta than Toronto?

6. Mercury is 57.6 million kilometers from the Sun.
   Earth is 148.64 million kilometers from the Sun.

   How much farther from the Sun is the Earth?

7. Continue the patterns.   a) .2,   .4,   .6, ____, ____, ____   b) .3,   .6,   .9, ____, ____, ____

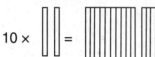

If a hundreds block represents 1 whole,
then a tens block represents 1 tenth (or 0.1).

and

$10 \times$ ▯ = ▯▯▯▯

10 tenths make 1 whole:
$10 \times 0.1 = 1.0$

---

1. Multiply the number of tens blocks by 10. Then show how many hundreds blocks you would have.
   The first one is done for you.

   a)
   $10 \times$ ▯▯ = ▯▯▯▯▯

   b)
   $10 \times$ ▯▯▯ =

   $10 \times 0.2 =$ __2__

   $10 \times 0.3 =$ _____

   c)
   $10 \times$ ▯▯▯▯▯▯ =

   $10 \times 0.6 =$ _____

2. Multiply.

   a) $10 \times .5 =$ ____
   b) $10 \times .7 =$ ____
   c) $10 \times 1.4 =$ ____
   d) $10 \times .9 =$ ____

   e) $10 \times 1.7 =$ ____
   f) $1.6 \times 10 =$ ____
   g) $18.2 \times 10 =$ ____
   h) $17.3 \times 10 =$ ____

   i) $10 \times 23.5 =$ ____
   j) $10 \times 1.72 =$ ____
   k) $10 \times 42.6 =$ ____
   l) $5.36 \times 10 =$ ____

3. To change from dm to cm, you multiply
   by 10 (there are 10 cm in 1 dm).

   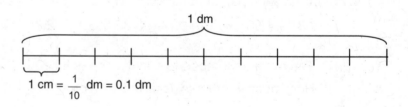

   1 dm

   $1 \text{ cm} = \frac{1}{10} \text{ dm} = 0.1 \text{ dm}$

   Find the answers.

   a) $.6 \text{ dm} =$ _____ cm
   b) $.8 \text{ dm} =$ _____ cm
   c) $1.6 \text{ dm} =$ _____ cm

4. 10×3 can be written as a sum: $3 + 3 + 3 + 3 + 3 + 3 + 3 + 3 + 3 + 3$.
   Write $10 \times .3$ as a sum and skip count by .3 to find the answer.

5. A dime is a tenth of a dollar (10¢ = $0.10).
   Draw a picture or use play money to show that $10 \times \$0.20 = \$2.00$.

NS6-87: Multiplying Decimals by 100 and 1000

 = 1.0   □ = 0.01      100 × □ =

If a hundreds block represents 1 whole then
a ones block represents 1 hundredth (or .01).

100 hundredths makes 1 whole:
100 × .01 = 1.00

---

1. Write a multiplication statement for each picture.

    a) 100 × □ =

    _____100 × .02_____ = _____

    b) 100 × □ =

    _____ = _____

2. The picture shows why the decimal shifts two places to the right when multiplying by 100.

    100 ×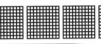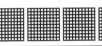

    = 100 × 0.12 = ___12___   100 × 0.1 = ___10___   100 × 0.02 = ___2___

    In each case shift the decimal 2 places to the right.

    a) 100 × .7 = ___70___   b) 100 × 1.8 = _____   c) 100 × 4.6 = _____

    d) 100 × 5.9 = _____   e) 100 × 2.3 = _____   f) 100 × 4.0 = _____

    g) 100 × 0.16 = _____   h) 100 × 0.69 = _____   i) 100 × 0.07 = _____

3. Multiply.

    a) 100 × .07 = ___7___   b) 100 × .06 = _____   c) 100 × .67 = _____   d) .95 × 100 = _____

    e) 100 × 1.82 = _____ f) 100 × 4.07 = _____ g) 100 × .50 = _____   h) 100 × .7 = _____

    i) 100 × 1.8 = _____   j) 100 × .35 = _____   k) 100 × .64 = _____   l) .95 × 100 = _____

4. a) What do 1000 thousandths add up to? _____   b) What is 1000 × .001? _____

5. Look at your answer to Question 4 b).

    How many places right does the decimal shift when you multiply by 1000? _____

6. Multiply the numbers by shifting the decimal.

    a) 1000 × .86 = _____   b) 1000 × .325 = _____   c) 1000 × 1.329 = _____

    d) 1000 × .76 = _____   e) 1000 × 8.25 = _____   f) 1000 × 7.5 = _____

Number Sense 2

NS6-88: Multiplying Decimals by Whole Numbers

The picture shows how to multiply a decimal by a whole number.

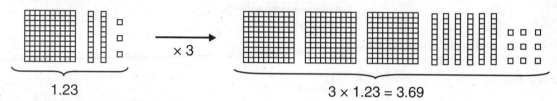

1.23 $\times 3$ $3 \times 1.23 = 3.69$

HINT: Simply multiply each digit separately.

----

1. Multiply mentally.

    a) $2 \times 1.43 =$ _____   b) $3 \times 1.2 =$ _____   c) $5 \times 1.01 =$ _____   d) $4 \times 2.1 =$ _____

    e) $2 \times 5.34 =$ _____   f) $4 \times 2.1 =$ _____   g) $3 \times 3.12 =$ _____   h) $3 \times 4.32 =$ _____

2. Multiply by exchanging tenths for ones (the first one is done for you).

    a) 6×1.4 = __$6 \times 1 = 6$__ ones + __$6 \times 4 = 24$__ tenths = __8__ ones + __4__ tenths = __8.4__

    b) 3×2.5 = _____ ones + _____ tenths = _____ ones + _____ tenths = _____

    c) 3×2.7 = _____ ones + _____ tenths = _____ ones + _____ tenths = _____

    d) 4×2.6 = _____

3. Multiply by exchanging tenths for ones or hundredths for tenths.

    a) $3 \times 2.51 =$ _____ ones + _____ tenths + _____ hundredths

    = _____ ones + _____ tenths + _____ hundredths = _____

    b) $4 \times 2.14 =$ _____ ones + _____ tenths + _____ hundredths

    = _____ ones + _____ tenths + _____ hundredths = _____

    c) $5 \times 1.41 =$ _____ ones + _____ tenths + _____ hundredths

    = _____ ones + _____ tenths + _____ hundredths = _____

4. Multiply. In some questions you will have to regroup twice.

    a) 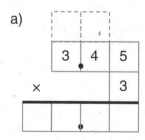   b) 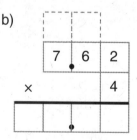   c) 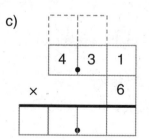   d)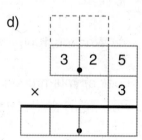

5. In your notebook, find the products.

    a) 5×2.1 b) 3×8.3 c) 5×7.5 d) 9×2.81 e) 7×3.6 f) 6×3.4

    g) 4×3.2 h) 5×6.35 i) 6×3.95 j) 8×2.63 k) 3×31.21 l) 4×12.32

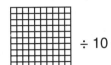 ÷ 10 = ÷ 10 = □ 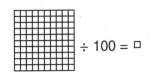 ÷ 100 = □

Divide 1 whole into
10 equal parts –
each part is 1 tenth:
1.0 ÷ 10 = 0.1

Divide 1 tenth into
10 equal parts –
each part is 1 hundredth:
0.1 ÷ 10 = 0.01

Divide 1 whole into
100 equal parts –
each part is 1 hundredth:
1.0 ÷ 100 = 0.01

When you divide a decimal by 10 the
decimal shifts <u>one place to the left</u>:

0.7 ÷ 10 = .07   7.0 ÷ 10 = .7

When you divide a decimal by 100 the
decimal shifts <u>two places to the left.</u>

7.0 ÷ 100 = .07

- -

1. Complete the picture and write a division statement for each picture.

a)  ÷ 10 =   b) ÷ 10 =

   $\underline{\quad 2.0 \div 10 \quad} = \underline{\quad .2 \quad}$ $\underline{\qquad\qquad} = \underline{\qquad}$

c) 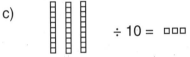 ÷ 10 = □□□   d) ÷ 10 =   e) ÷ 10 =

   $\underline{\quad .3 \div 10 \quad} = \underline{\qquad}$ $\underline{\qquad\qquad} = \underline{\qquad}$ $\underline{\qquad\qquad} = \underline{\qquad}$

2. Complete the picture and write a division statement (the first one is done for you).

a) ÷ 10 =   b) ÷ 10 =

   $\underline{\quad 2.3 \div 10 \quad} = \underline{\quad .23 \quad}$ $\underline{\qquad\qquad} = \underline{\qquad}$

3. Shift the decimal one or two places to the left by drawing an arrow as shown in 3 a).
   HINT: If there is no decimal, add one to the right of the number first.

   a) 0.3 ÷ 10 = $\underline{\quad.03\quad}$ b) 0.5 ÷ 10 = _____   c) 0.7 ÷ 10 = _____   d) 1.3 ÷ 10 = _____

   e) 7.6 ÷ 10 = _____   f) 12.0 ÷ 10 = _____   g) 9 ÷ 10 = _____   h) 6 ÷ 10 = _____

   i) 42 ÷ 10 = _____   j) 17 ÷ 10 = _____   k) .9 ÷ 10 = _____   l) 27.3 ÷ 10 = _____

   m) 3.0 ÷ 100 = _____   n) 6.2 ÷ 100 = _____   o) .7 ÷ 100 = _____   p) 17.2 ÷ 100 = _____

4. Explain why 1.00 ÷ 100 = .01 using dollar coins as a whole.

5. A wall 3.5 m wide is painted with 100 stripes of equal width.
   How wide is each stripe?

6. 5 × 3 = 15 and 15 ÷ 5 = 3 are in the same fact family.
   Write a division statement in the same fact family as 10 × 0.1 = 1.0.

You can divide a decimal by a whole number by making a base ten model. Keep track of your work using long division. Use a hundreds block to represent 1 whole, a tens block to represent 1 tenth and a ones block to represent 1 hundredth.

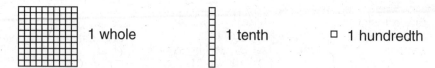

1 whole   1 tenth   □ 1 hundredth

- -

1. Find **5.12 ÷ 2** by drawing a base ten model and by long division.

Step 1: Draw a base ten model of 5.12.

> Draw your model here:

Step 2: Divide the whole blocks into 2 equal groups.

remaining wholes, tenths and hundredths

Step 3: Exchange the left over whole blocks for 10 tenths.

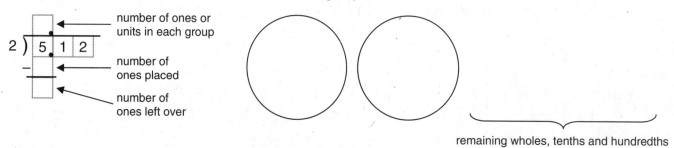

regroup a whole as 10 tenths

REMEMBER: A whole is represented by a hundreds block.

Step 4: Divide the tenths blocks into 2 equal groups.

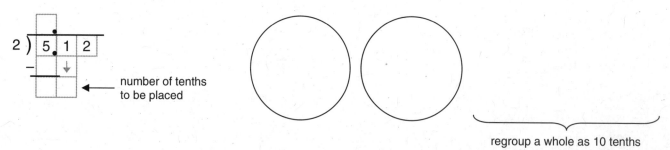

remaining tenths and hundredths

jump math
MULTIPLYING POTENTIAL

Number Sense 2

<u>Step 5:</u>   Regroup the left over tenth block as 10 hundredths.

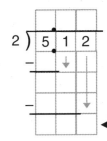

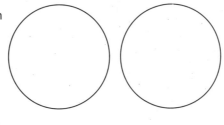

← number of hundredths
   to be placed

exchange a tenth for 10 hundredths

<u>Steps 6 and 7:</u>   Divide the hundredths into 2 equal groups.

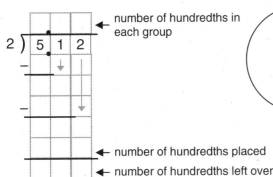

← number of hundredths in
   each group

← number of hundredths placed
← number of hundredths left over

remaining hundredths

2. Divide.

a) 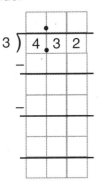   b) 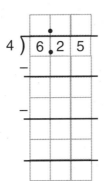   c) 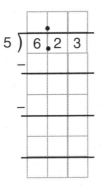   d)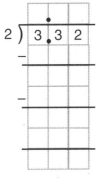

3. Divide.   a) $8\overline{)1.44}$ b) $7\overline{)9.4}$ c) $8\overline{)2.72}$ d) $9\overline{)6.13}$ e) $5\overline{)20.5}$

4. Five apples cost $2.75. How much does each apple cost?

5. Karen cycled 62.4 km in 4 hours.
    How many kilometres did she cycle in an hour?

6. Four friends earn a total of $29.16 shovelling snow.
    How much does each friend earn?

7. Which is a better deal: 6 pens for $4.98 or 8 pens for $6.96?

8. James divides 3.4 m of rope into 6 equal parts. Each part is a whole number of decimetres long.

    a) How long is each part?   b) How many decimetres of rope are left over?

1. Fill in the blanks.

   a) .74 + .1 = _____

   b) .23 + .1 = _____

   c) .09   + .1 = _____

   d) .79 + .1 = _____

   e) .50 + .01 = _____

   f) 2.79 + .01 = _____

   g) 3.056 + .001 = _____

   h) 0.009 + .001 = _____

   i) 2.372 + .01 = _____

2. Fill in the blanks.

   a) _____ is .1 more than .4

   b) _____ is .1 more than 0.9

   c) _____ is .1 more than 3.25

   d) _____ is .01 more than .75

   e) _____ is .01 more than .79

   f) _____ is .001 more than .372

3. Fill in the blanks.

   a) 2.34 + _____ = 2.35

   b) 3.75 + _____ = 3.85

   c) 8.07 − _____ = 8.06

   d) 6.92 − _____ = 6.82

   e) 3.957 + _____ = 3.967

   f) 7.852 + _____ = 7.853

4. Fill in the missing numbers on the number lines.

   a)
   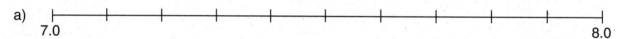
   7.0   8.0

   b)
   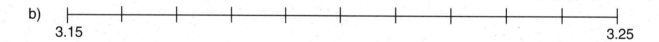
   3.15   3.25

   c)
   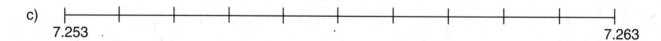
   7.253   7.263

5. Continue the patterns.

   a) .3, .4, .5, _____, _____, _____

   b) 9.6, 9.7, 9.8 , _____, _____, _____

   c) 2.5, 2.6, 2.7, _____, _____, _____

   d) 4.34, 4.35, 4.36 _____, _____, _____

   e) 2.96, 2.97, 2.98, _____, _____, _____

   f) 1.234, 1.235 , _____, _____

6. Fill in the blanks.

   a) 7.9 + .1 = _____

   b) 2.9 + .1 = _____

   c) 6.95 + .1 = _____

   d) 2.69 + .01 = _____

   e) 3.99 + .01 = _____

   f) 7.299 + .001 = _____

NS6-92: Place Value and Rounding

1. Round to the nearest whole number.

   a) 26.408   b) 38.97   c) 59.962   d) 71.001

2. Round to the nearest tenth.

   a) 26.54   b) 47.946   c) 49.874   d) 38.96

3. Estimate by rounding to the nearest whole number.

   a) 94.7 ÷ 5.2   b) 2.96 × 8.147   c) 4.51 × 0.86

   Estimate.   Estimate.   Estimate.

4. Estimate by rounding each number to the nearest whole number. Use your estimate to say which answers are reasonable.

   a) 32.7 + 4.16 = 73.8   b) 0.7 × 8.3 = 5.81   c) 9.2 × 10.3 = 947.6

   d) 97.2 ÷ 0.9 = .8   e) 88.2 ÷ 9.8 = 9   f) 54.3 − 18.6 = 35.7

5. Calculate each answer from Question 4. Were your predictions correct?

6. 5.3 is said to be precise to the tenths. Any number from 5.25 to 5.34 could round it.
   What numbers could round to 7.2?

7. Which measurement would need to be taken to the nearest tenth?

   a) Height of a building (metres).
   b) Distance shot put is thrown in Olympic Games (metres).
   c) Length of the Canada/U.S. border (kilometres).

8. What amount is represented by the tenths digit?

   a) 3.54 m   b) 6.207 km   c) 4.69 dm

   d) 4.6 million   e) $17.46   f) 83.4 cm

jump math
MULTIPLYING POTENTIAL.

Number Sense 2

The size of a unit of measurement depends on which unit has been selected as the <u>whole</u>.

A millimetre is a **tenth** of a centimetre,
but it is only a **hundredth** of a decimetre.
REMEMBER: A decimetre is 10 centimetres.

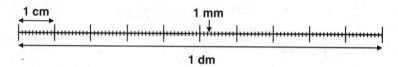

---

1. Draw a picture in the space provided to show 1 tenth of each whole.

   a)

   1 whole   1 tenth

   b)

   1 whole   1 tenth

   c)

   1 whole   1 tenth

2. Write each measurement as a fraction then as a decimal.

   a) 1 cm = $\dfrac{1}{10}$ dm = __.1__ dm

   b) 100 mm = ☐ m = _____ m

   c) 1 mm = ☐ cm = _____ cm

   d) 16 mm = ☐ cm = _____ .cm

   e) 77 mm = ☐ dm = _____ dm

   f) 83 cm = ☐ m = _____ m

3. Add the measurements by first changing the <u>smaller unit</u> into a decimal in the <u>larger unit</u>.

   a) 5 cm + 7.3 dm = __0.5 dm + 7.3 dm = 7.8 dm__

   b) 5 cm + 3.2 dm = _____

   c) 8 mm + 5.7 cm = _____

   d) 33 cm + 1.64 m = _____

   e) 685 cm + 12.3 m = _____

   f) 982 cm + 1.5 m = _____

4. Write a decimal for each description. Some questions have more than one answer.

   a) Between 4.31 and 4.34   ☐.☐☐

   b) Between 2.60 and 2.70   ☐.☐☐

   c) Between 13.75 and 13.8   ☐☐.☐☐

   d) Between 9.7 and 9.8   ☐☐.☐☐

   e) One tenth greater than 5.23   ☐.☐☐

   f) One hundredth less than 4.00   ☐.☐☐

5. Add.

   a) $4\,000 + 300 + 7 + 0.01 =$ _____

   b) $20\,000 + 300 + 30 + 0.2 + .04 =$ _____

   c) $300\,000 + 20\,000 + 5000 + 70 + 0.1 + 0.09 + 0.006 =$ _____

6. Write < or > to show which decimal is greater.

   a) 4.9 ☐ 4.6   b) 3.45 ☐ 3.35   c) 1.9 ☐ 1.26   d) 0.7 ☐ 0.524

7. Put a decimal in each number so that the digit **7** has the value $\frac{7}{10}$.

   a) 5 7 2   b) 1 0 7   c) 2 8 7 5 9   d) 7

8. Use the digits 5, 6, 7 and 0 to write a number between the given numbers.

   a) .567 < _____ < .576   b) 5.607 < _____ < 5.760

9. Write three decimals between .3 and .5: _____ _____ _____

10. Write −, +, ×, or ÷ in the circle.

   a) 62.57 ◯ 10 = 72.57   b) 19.2 ◯ 10 = 192   c) 9 ◯ 10 = .9

11. Write the decimals in order from least to greatest. Explain your answer for c).

   a) .37   .275   .371   b) .007   .07   .7   c) 1.29   1.3   2.001

12. Use the number line to estimate which fraction lies in each range.

   Fractions: $\frac{1}{2}$, $\frac{1}{3}$, $\frac{3}{4}$, $\frac{1}{10}$ Ranges:

A	B	C	D	E
0 to .2	.2 to .4	.4 to .6	.6 to .8	.8 to 1.0

13. Is 6 a reasonable estimate for 8 × .72? Explain.

14. How do you know that 10 × 87.3 is 873 and not 8 730?

15. Change 1.25 hours to a mixed fraction in lowest terms, then to a time in minutes.

Answer the following questions in your notebook.

1. Explain how you would change 5.47 m into cm.

   HINT: How many centimeters are in $\frac{47}{100}$ of a metre?

2. $0.68 means 6 dimes and 8 pennies.

   Why do we use the decimal notation for money?

   What is a dime a tenth of?

   What is a penny a hundredth of?

3. The wind speed in Vancouver was 26.7 km/h on Monday, 16.0 km/h on Tuesday and 2.4 km/h on Wednesday.

   What was the average wind speed over the 3 days?

4. 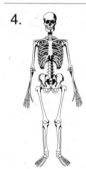 The tallest human skeleton is 2.7 m high and the shortest is 60 cm high. What is the difference in the heights of the skeletons?

5.

Star	Distance from the Sun in light years
Alpha Centauri	4.3
Barnard's Star	6.0
L726-8	8.8
Sirius	9.5
61 Cigni	11.0

A light year is 9.5 trillion kilometers.

NOTE: This is the distance light travels in one year.

a) If you traveled from the Sun to Barnard's Star, how many trillion km would you have traveled?

b) Which star is just over twice as far from the Sun as Alpha Centauri?

6. Food moves through the esophagus at a rate of .72 km per hour.

   How many meters per hour is this?

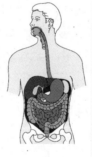

7. Write the following prices in order from least to greatest.
   What is the difference between the highest and the lowest prices?
   HINT: Change all the prices to dollars per kilogram.

 A: Cherries – 59¢ for each 100 g

 **B:** Watermelon – $3.90 for each kg

 C: Strawberries – $0.32 for each 100 g

 **D:** Blueberries – $3.99 for each 500 g

NS6-95: Unit Rates

A rate is a comparison of two quantities in different units.

In a **unit rate**, one of the quantities is equal to one.

For instance, "1 apple costs 30¢" is a unit rate.

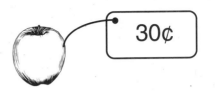

30¢

---

1. Fill in the missing information.

   a) 17 km in 1 hour

   _____ km in 3 hours

   b) 1 book costs $4.95

   4 book costs _____

   c) 2 teachers for 75 students

   6 teachers for _____

   d) 5 mangoes cost $4.95

   1 mango costs _____

   e) 4 mangoes cost $12

   1 mango costs _____

   f) 6 pears cost $4.92

   1 pear costs _____

2. Find the unit rate.

   a) 4 kg of rice for 60 cups of water.

   1 kg of rice for _____ cups of water.

   b) 236 km in 4 hours.

   _____ km in 1 hr.

   c) 6 cans of juice cost $1.98

   1 can costs _____

3. Use a ruler to find out the height of each animal. (One centimetre represents 50 cm in real life.)

   a)
   Kangaroo

   Height of picture in cm: _____

   Height of animal in m: _____

   b)

   Height of picture in cm: _____

   Height of animal in m: _____

   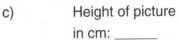

   c)

   Height of picture in cm: _____

   Height of animal in m: _____

   Horse

   Giraffe

4. Ron earns $66 babysitting for six hours.
   How much does he earn in an hour?

   _____

5. Tina earns $75 cutting lawns for 5 hours.
   How much does she earn in an hour?

   _____

NS6-96: Introduction to Ratios

A **ratio** is a comparison between two numbers.

1.

   a) The ratio of moons to circles is _____ : _____

   b) The ratio of triangles to moons is _____ : _____

   c) The ratio of cylinders to squares is _____ : _____

   d) The ratio of squares to circles is _____ : _____

   e) The ratio of squares to moons is _____ : _____

   f) The ratio of squares to figures is _____ : _____

2. Write the number of vowels compared to the number of consonants in the following words.

   a) apple   __2__ : __3__

   b) banana   ____ : ____

   c) orange   ____ : ____

   d) pear   ____ : ____

3. Write the ratio of the lengths.

   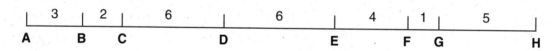

   a) AB to DE _____ : _____

   b) BC to CD _____ : _____

   c) EF to FG _____ : _____

   d) EF to BC _____ : _____

   e) AB to GH _____ : _____

   f) CD to FG _____ : _____

4. To make punch, you need …

   - 4 L of ginger ale
   - 2 L of orange juice
   - 3 L of mango juice

   What is the ratio of ginger ale to punch?

   _____

5.

   a) In the above pattern, what does the ratio 2 : 3 describe?

   _____

   b) What does the ratio 5 : 10 describe?

   _____

6. Build a model or draw a picture that could be described by the ratio 3 : 4.

NS6-97: Equivalent Ratios

1. The picture shows that the ratio of apples to bananas on a grocery shelf is:

   8 apples to every 6 bananas

   OR

   4 apples to every 3 bananas.

   Group the fruit to show two equivalent ratios.

   a)

   _____ to every _____

   or _____ to every _____

   b)

   _____ to every _____

   or _____ to every _____

2. Starting with the ratio 2 triangles to every 3 squares, Talia created a sequence of equivalent ratios.
   Fill in the missing figures and ratios.

Triangles	△△	△△ △△	△△ △△ △△	
Squares	☐☐☐	☐☐☐ ☐☐☐		☐☐☐ ☐☐☐ ☐☐☐ ☐☐☐
Ratio	2 : 3			

3. Starting with the ratio given, write a sequence of five ratios that are all equivalent.

   a) $3 : 4 = 6 : 8 = \quad : \quad = \quad : \quad = \quad :$

   b) $2 : 5 =$

4. Find the missing terms.   a) $3 : 4 = \underline{\quad} : 8$ b) $5 : 7 = 10 : \underline{\quad}$ c) $2 : 5 = \underline{\quad} : 25$

A recipe for granola calls for 2 cups of raisins for every 3 cups of oats.

How many cups of raisins will Eschi need for 12 cups of oats? $2 : 3 = 4 : 6 = 6 : 9 = 8 : 12$

She writes a sequence of equivalent ratios to find out.

NOTE: She multiplies both terms in the ratio 2 : 3 by 2, then by 3, then by 4.   Eschi needs 8 cups of raisins.

---

5. Solve each problem by writing a sequence of equivalent ratios (as in the example above).

   a) A recipe calls for 5 cups of oats for every 3 cups of raisins.
      How many cups of oats are needed for 12 cups of raisins?

   b) 2 cm on a map represent 11 km.
      How many km do 8 cm on the map represent?

   c) Six bus tickets cost $5.
      How much will 18 tickets cost?

NS6-98: Finding Equivalent Ratios

> There are 3 boys for every 2 girls in a class of 20 children.
>
> To find out how many boys are in the class, write a sequence of ratios.
>
> 3 boys : 2 girls = 6 boys : 4 girls = 9 boys : 6 girls = 12 boys : 8 girls
>
> Stop when the terms of the ratio add to 20.
>
> 12 boys + 8 girls = 20 students. So there are 12 boys in the class.

1. Write a sequence of ratios to solve each problem. The first one is started for you.

   a) There are 5 boys for every 4 girls in a class of 27 children.
      How many girls are in the class?

      $5 : 4$ = $10 : 8$ =

   b) There are 3 red fish for every 5 blue fish in an aquarium.
      With 24 fish, how many fish are blue?

   c) A recipe for punch calls for 3 L of orange juice for every 2 litres of mango juice.
      How many litres of orange juice are needed to make 15 litres of punch?

> Five subway tickets cost $4. Kyle wants to know how much 20 tickets will cost. He writes the ratio of tickets to dollars as a fraction. Then, he finds an equivalent fraction by multiplying:
>
> Step 1: $\dfrac{4}{5} = \dfrac{?}{20}$
>
> Step 2: $\dfrac{4}{5} = \dfrac{\ \ }{20}$
>
> Step 3: $\dfrac{4}{5} \underset{\times 4}{\overset{\times 4}{=}} \dfrac{\ \ }{20}$
>
> Step 4: $\dfrac{4}{5} \underset{\times 4}{\overset{\times 4}{=}} \dfrac{16}{20}$

2. Solve the following ratios. Draw arrows to show what you multiply by.

   a) $\dfrac{3}{4} \underset{\times 5}{\overset{\times 5}{=}} \dfrac{\ \ }{20}$

   b) $\dfrac{1}{5} = \dfrac{\ \ }{25}$

   c) $\dfrac{2}{5} = \dfrac{\ \ }{20}$

   d) $\dfrac{6}{7} = \dfrac{\ \ }{35}$

   e) $\dfrac{3}{4} = \dfrac{\ \ }{16}$

   f) $\dfrac{2}{3} = \dfrac{\ \ }{12}$

   g) $\dfrac{15}{25} = \dfrac{\ \ }{100}$

   h) $\dfrac{5}{9} = \dfrac{\ \ }{45}$

BONUS

NOTE: Sometimes, in the questions below, the arrow may point from right to left.

3. a) $\dfrac{15}{\ \ } \underset{\times 5}{\overset{\times 5}{=}} \dfrac{3}{4}$

   b) $\dfrac{10}{\ \ } = \dfrac{2}{5}$

   c) $\dfrac{9}{\ \ } = \dfrac{3}{7}$

   d) $\dfrac{10}{15} = \dfrac{\ \ }{3}$

   e) $\dfrac{4}{5} = \dfrac{\ \ }{15}$

   f) $\dfrac{2}{3} = \dfrac{\ \ }{9}$

   g) $\dfrac{\ \ }{45} = \dfrac{2}{5}$

   h) $\dfrac{\ \ }{20} = \dfrac{7}{10}$

NS6-99: Word Problems (Advanced)

In a pet shop, there are 3 cats for every 2 dogs. If there are 12 cats in the shop, how many dogs are there?

Solution:

Step 1:
Write, as a fraction, the ratio of the two things being compared.

$$\frac{3}{2}$$

Step 2:
Write, in words, what each number stands for.

cats $\frac{3}{2}$
dogs

Step 3:
On the other side of an equals sign, write the same words, on the same levels.

cats $\frac{3}{2}$ = —— cats
dogs   dogs

Step 4:
Re-read the question to determine which quantity (i.e. number of cats or dogs) has been given (in this case, cats) – then place that quantity on the proper level.

cats $\frac{3}{2}$ = $\frac{12}{\quad}$ cats
dogs   dogs

Step 5:
Solve the ratio.

Solve the following questions in your notebook.

1. There are 2 apples in a bowl for every 3 oranges.
   If there are 9 oranges, how many apples are there?

2. Five bus tickets costs $3.
   How many bus tickets can you buy with $9?

3. A basketball team won 2 out of every 3 games they played. They played a total of 15 games.
   How many games did they win?
   NOTE: The quantities are "games won" and "games played."

4. To make fruit punch, you mix 1 litre of orange juice with 2 litres of pineapple juice.
   If you have 3 litres of orange juice, how many litres of pineapple juice do you need?

5. Nora can run 3 laps in 4 minutes.
   At that rate, how many laps could she run in 12 minutes?

6. The ratio of boys to girls in a class is 4:5.
   If there are 20 boys, how many girls are there?

7. 2 cm on a map represents 5 km in real life.
   If a lake is 6 cm long on the map, what is its actual size?

Tony can paint 3 walls in $\frac{1}{2}$ an hour. He wants to know how many walls he can paint in 5 hours.

He first changes the ratio $\frac{1}{2}$: 3 to a more convenient form by doubling both terms of the ratio.

$\frac{1}{2}$ hour : 3 walls = 1 hour : 6 walls

Then he multiplies each term by 5.

1 hour : 6 walls = 5 hours : 30 walls

Tony can paint 30 walls in 5 hours.

---

1. Change each ratio so the number on the left is a whole number.

a) $\frac{1}{2}$ hour : 2 km walked =

b) $\frac{1}{4}$ cup of flour : 2 cups of potatoes =

c) $\frac{1}{3}$ hour : 3 km rowed =

d) $\frac{1}{3}$ cup of raisins : 2 cups of oats =

e)  0.3 km : 1 litre of gas used =

f) 1.7 mL of ginger ale : 0.3 mL of orange juice =
HINT: For this ratio, multiply each term by 10.

2. Solve each problem by changing the ratio into a more convenient form.

a) Rhonda can ride her bike 3 km in $\frac{1}{4}$ of an hour. How far can she ride in 2 hours?

b) A plant grows 3 cm in 4 days. How many days will it take to grow 6 cm?

3.  How many equivalent ratios can you write for this array?

4. For each ratio below write an equivalent ratio with one term equal to 20.
    a) 4 : 6   b) 3 : 5   c) 4 : 5   d) 10 : 30

5. In a class of 30 students there are 10 girls. Explain why the ratio of girls to boys is 1 to 2.

NS6-101: Percents

A **percent** is a ratio that compares a number to 100.

The term percent means "out of 100" or "for every 100." For instance, 84% on a test means 84 out of 100.

You can think of percent as a short form for a fraction with 100 in the denominator, e.g. $45\% = \frac{45}{100}$

1. Write the following percents as fractions.

   a) 7%   b) 92%   c) 5%   d) 15%

   e) 50%   f) 100%   g) 2%   h) 7%

2. Write the following fractions as percents.

   a) $\frac{2}{100}$ b) $\frac{31}{100}$ c) $\frac{52}{100}$ d) $\frac{100}{100}$

   e) $\frac{17}{100}$ f) $\frac{88}{100}$ g) $\frac{2}{100}$ h) $\frac{1}{100}$

3. Write the following decimals as percents, by first turning them into fractions. The first one has been done for you.

   a) $.72 = \frac{72}{100} = 72\%$ b) $.27$ c) $.04$

4. Write the fraction as a percent by changing it to a fraction over 100. The first one has been done for you.

   a) $\dfrac{3 \,^{\times 20}}{5 \,_{\times 20}} = \dfrac{60}{100} = 60\%$ b) $\frac{2}{5}$

   c) $\frac{4}{5}$ d) $\frac{1}{4}$

   e) $\frac{3}{4}$ f) $\frac{1}{2}$

   g) $\frac{3}{10}$ h) $\frac{7}{10}$

   i) $\frac{17}{25}$ j) $\frac{7}{20}$

   k) $\frac{3}{25}$ l) $\frac{19}{20}$

   m) $\frac{23}{50}$ n) $\frac{47}{50}$

NS6-101: Percents (continued)

5. Write the following decimals as a percents. The first one has been done for you.

   a) .2 $= \dfrac{2}{10} \stackrel{\times 10}{\underset{\times 10}{=}} \dfrac{20}{100} = 20\%$

   b) .5

   c) .7

   d) .9

6. What percent of the figure is shaded?

   a)

   b)

   c)

   d)

7. Change the following fractions to percents by first reducing them to lowest terms.

   a) $\dfrac{9}{15} \stackrel{\div 3}{\underset{\div 3}{=}} \dfrac{3}{5} = \dfrac{3}{5} \stackrel{\times 20}{\underset{\times 20}{=}} \dfrac{60}{100} = 60\%$

   b) $\dfrac{12}{15}$

   c) $\dfrac{3}{6}$

   d) $\dfrac{7}{35}$

   e) $\dfrac{21}{28}$

   f) $\dfrac{18}{45}$

   g) $\dfrac{12}{30}$

   h) $\dfrac{10}{40}$

   i) $\dfrac{20}{40}$

   j) $\dfrac{16}{40}$

   k) $\dfrac{60}{150}$

   l) $\dfrac{45}{75}$

NS6-102: Visual Representations of Percents

1. Fill in the chart below. The first one has been done for you.

Drawing				
Fraction	$\frac{23}{100}$	$\frac{}{100}$	$\frac{45}{100}$	$\frac{}{100}$
Decimal	0._2_3_	0.___ ___	0.___ ___	0.81
Percent	23%	63%	_____ %	_____ %

Use a ruler for Questions 2 to 5.

2. Shade 50% of each box.

   a)   b)   c)

3. Shade 25% of each box.

   a)    b)

4. The triangle is 50% of a parallelogram. Show what 100% might look like.

   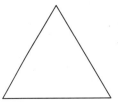

5. Colour 50% of the rectangle blue, 40% red, and 10% green.

6. a) Write a fraction for the part shaded: _____

   b) Write the fraction with a denominator of 100: _____

   c) Write a decimal and percent for the part shaded: _____ _____

   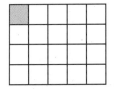

Number Sense 2

7. Write a fraction and a percent for each division of the number line.

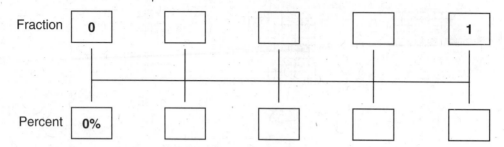

Fraction | 0 | | | | 1

Percent | 0% | | | |

8. Draw marks to show 25%, 50% and 75% of the line segment.

a) ——————————————   b) ————————————————————————

c) ————————————   d) ————————————————————————————

9. Extend each line segment to show 100%.

a) ├——— 50% ———┤   b) ├ 25% ┤

c) ├ 20% ┤   d) ├——— 75% ———┤

e) ├——————————┤
   0%   60%

f) ├——————————————┤
   0%   80%

g) ├——┤
   0% 10%

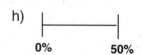
h) ├——————┤
   0%   50%

10. Estimate the percent of the line segment represented by each mark.

a) ├——————×——————┤
   0%   100%

b) ├———×————————┤
   0%   100%

11. Draw a rough sketch of a floor plan for a museum.

   The different collections should take up the following amounts of space:

   • Dinosaurs   40%
   • Animals   20%
   • Rocks and Minerals 10%
   • Ancient Artifacts   20%

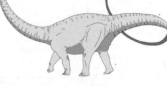

   Washrooms should take up the final 10% of the floor space.

12. Asia covers 30% of the world's land mass.
   Using a globe, compare the size of Asia to the size of Australia.
   Approximately what percent of the world's land mass does Australia cover?

NS6-103: Comparing Decimals, Fractions & Percents

1. From the list below, choose the percent to which each fraction is closest.

10%	25%	50%	75%	100%

a) $\frac{3}{5}$ _____

b) $\frac{4}{5}$ _____

c) $\frac{2}{5}$ _____

d) $\frac{2}{10}$ _____

e) $\frac{1}{10}$ _____

f) $\frac{4}{10}$ _____

g) $\frac{9}{10}$ _____

h) $\frac{4}{25}$ _____

i) $\frac{11}{20}$ _____

j) $\frac{16}{20}$ _____

k) $\frac{37}{40}$ _____

l) $\frac{1}{12}$ _____

2. Write <, > or = between the following pairs of numbers. The first one has been done for you.
   HINT: Change each pair of numbers to a pair of fractions with the same denominator.

a) $\frac{1}{2}$ ☐ 47%

$\frac{50 \times 1}{50 \times 2}$ ☐ $\frac{47}{100}$

$\frac{50}{100}$ > $\frac{47}{100}$

b) $\frac{1}{2}$ ☐ 53%

c) $\frac{1}{4}$ ☐ 23%

d) $\frac{3}{4}$ ☐ 70%

e) $\frac{2}{5}$ ☐ 32%

f) .27 ☐ 62%

g) .02 ☐ 11%

h) $\frac{1}{10}$ ☐ 10%

i) $\frac{19}{25}$ ☐ 93%

j) $\frac{23}{50}$ ☐ 46%

k) .9 ☐ 10%

l) $\frac{11}{20}$ ☐ 19%

3. Write each set of numbers in order from least to greatest by first changing each number to a <u>fraction</u>.

a) $\frac{3}{5}$, 42% , .73

b) $\frac{1}{2}$, .73 , 80%

c) $\frac{1}{4}$, .09 , 15%

d) $\frac{2}{3}$, 57% , .62

Number Sense 2

NS6-104: Finding Percents

If you use a thousands cube to represent 1 whole, you can see that taking $\frac{1}{10}$ of a number is the same as dividing the number by 10 – the decimal shifts one place left.

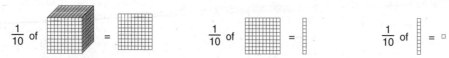

$\frac{1}{10}$ of 1 = .1 $\frac{1}{10}$ of .1 = .01 $\frac{1}{10}$ of .01 = .001

---

1. Find $\frac{1}{10}$ of the following numbers by shifting the decimal.   Write your answers in the boxes provided.

   a) 4   b) 7   c) 32   d) 120   e) 3.8   f) 2.5

2. 10% is short for $\frac{1}{10}$. Find 10% of the following numbers.

   a) 9   b) 5.7   c) 4.05   d) 6.35   e) .06   f) 21.1

3. You can find percents that are multiples of 10 as follows.

   Example: Finding 30% of 21 is the same as finding 10% of 21
   and multiplying the result by 3.

   <u>Step 1:</u>   10% of 21 = $\boxed{2.1}$

   <u>Step 2:</u>   3 × $\boxed{2.1}$ = 6.3 → so 30% of 21 = 6.3

   Find the percents using the method above.

   a) 40% of 15

      i)   10% of 15 = $\boxed{}$

      ii) 4 × $\boxed{}$ = _____

   b) 60% of 25

      i)   10% of _____ = $\boxed{}$

      ii) _____ × $\boxed{}$ = _____

   c) 90% of 2.3

      i)   10% of _____ = $\boxed{}$

      ii) _____ × $\boxed{}$ = _____

   d) 60% of 35

      i)   10% of _____ = $\boxed{}$

      ii) _____ × $\boxed{}$ = _____

   e) 40% of 24

      i)   10% of _____ = $\boxed{}$

      ii) _____ × $\boxed{}$ = _____

   f) 20% of 1.3

      i)   10% of _____ = $\boxed{}$

      ii) _____ × $\boxed{}$ = _____

NS6-105: Finding Percents (Advanced)

35% is short for $\frac{35}{100}$. To find 35% of 27, Sadie finds $\frac{35}{100}$ of 27.

Step 1:   She multiplies 27 by 35.

2	3	
	2	7
×	3	5
1	3	5
8	1	0
9	4	5

Step 2:   She divides the result by 100.

945 ÷ 100 = 9.45

So 35% of 27 is 9.45.

- -

1. Find the following percents using Sadie's method.

a) 45% of 32

   Step 1:   Step 2:

   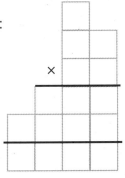

   ×

                                          _____ ÷ 100 =

                                          So _____ of _____ is _____.

b) 28% of 63

   Step 1:   Step 2:

   ×

                                          _____ ÷ 100 =

                                          So _____ of _____ is _____.

2. Find the following percents using Sadie's method.

   a) 13% of 9   b) 52% of 7   c) 65% of 8   d) 78% of 9

   e) 23% of 42   f) 17% of 68   g) 37% of 80   h) 62% of 75

3. 25% is equal to $\frac{1}{4}$ and 75% is equal to $\frac{3}{4}$. Find …

   a) 25% of 80   b) 25% of 280   c) 25% of 12   d) 75% of 20   e) 75% of 320

jump math
MULTIPLYING POTENTIAL

Number Sense 2

1. Find the missing percent of each child's stamp collection that comes from other countries.
   HINT: Change all fractions to percents.

a) Anne's collection:

Canada	USA	Other
40% = 40%	$\frac{1}{2}$ = 50%	 = 10%

b) Brian's collection:

Canada	England	Other
80%	$\frac{1}{10}$	

c) Juan's collection:

Mexico	USA	Other
$\frac{1}{2}$	40%	

d) Lanre's collection:

Canada	Nigeria	Other
22%	$\frac{3}{5}$	

e) Faith's collection:

Jamaica	Canada	Other
$\frac{3}{4}$	15%	

f) Carlo's collection:

France	Italy	Other
$\frac{3}{4}$	10%	

2. A painter spends $500.00 on art supplies. Complete the chart.

	Fraction of Money Spent	Percentage of Money Spent	Amount of Money Spent
Brushes			$50.00
Paint	$\frac{4}{10}$		
Canvas		50%	

3. Indra spent 1 hour doing homework. The chart shows the time she spent on each subject.

   a) Complete the chart.

   b) How did you find the amount of time spent on math?

Subject	Fraction of 1 hour	Percent of 1 hour	Decimal	Number of minutes
English	$\frac{1}{4}$		.25	15
Science	$\frac{1}{20}$	5%		
Math		50%		
French			.20	

4. Roger wants to buy a deck of cards that costs $8.00. The taxes are 15%. How much did he pay in taxes?

5. There are 15 boys and 12 girls in a class.
   $\frac{3}{4}$ of the girls have black hair, and 60% of the boys have black hair.
   How many children have black hair?

NS6-107: Circle Graphs

Many people use percents to show data.

Rita surveyed 100 Grade 6 students in her city about their favourite sport.

Favourite Sport	
baseball	32%
soccer	41%
hockey	16%
other	11%

She uses a circle divided into 100 equal parts to show her results.

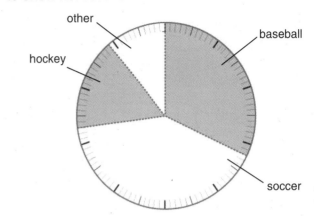

----

1. a) What percent of Grade 6 students like each sport in each city? Complete the chart.

City A

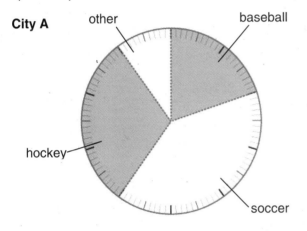

City B

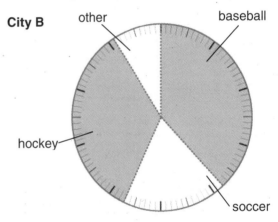

	baseball	soccer	hockey	other	total
City A					
City B					

 b) What is the total percent for each city? Why does this make sense?

2. Gisela copied down the following percents from a circle graph she saw on the Internet.

Favourite Sport			
baseball	soccer	hockey	other
48%	21%	26%	9%

How can you tell that she made a mistake?

Number Sense 2

3. Calli and Bilal go to different schools.
   They surveyed the Grade 6 students at their schools about their favourite subjects.

Calli's School	
Subject	**Number of students**
Science	10
Language	20
Gym	140
Other	30

Bilal's School	
Subject	**Number of students**
Science	20
Language	15
Gym	5
Other	10

a) How many Grade 6 students did Calli survey? _____ How many did Bilal survey? _____

b) Find the fraction of students in each school who like each subject.
   Turn the fraction into an equivalent fraction over 100, and then change it to a percent.

   Example: $\dfrac{\text{Number of students who like science in Calli's school}}{\text{Number of students in her school}} = \dfrac{10}{200} = \dfrac{5}{100} = 5\%$

c) Complete the circle graphs to show the percents you calculated in b).

Calli's School

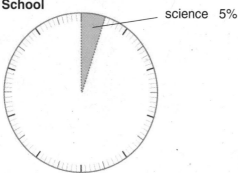

science 5%

Bilal's School

d) More people from Calli's school like Language than from Bilal's school. Why doesn't your circle graph show this?

4. About what percent of each circle is shaded?

a)

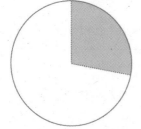

b)

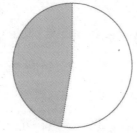

c)

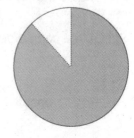

NS6-108: Fractions, Ratios and Percents

1. Write the number of girls (**g**), boys (**b**), and children (**c**) in each class.

   a) There are 8 boys and 5 girls in a class.   **b:** _____   **g:** _____   **c:** _____

   b) There are 4 boys and 7 girls in a class.   **b:** _____   **g:** _____   **c:** _____

   c) There are 12 boys and 15 girls in a class.   **b:** _____   **g:** _____   **c:** _____

   d) There are 9 girls in a class of 20 children. **b:** _____   **g:** _____   **c:** _____

   e) There are 7 boys in a class of 10 children.   **b:** _____   **g:** _____   **c:** _____

2. Write the number of boys, girls and children in each class.
   Then write the fraction of children who are boys and the fraction who are girls in the boxes provided.

   a) There are 5 boys and 6 girls in a class.   **b:** ____ ☐   **g:** ____ ☐   **c:** ____

   b) There are 15 children in the class. 8 are boys.   **b:** ____ ☐   **g:** ____ ☐   **c:** ____

3. Write the fraction of children in the class who are boys and the fraction who are girls.

   a) There are 5 boys and 12 children in the class.   **b:** ☐   **g:** ☐

   b) There are 3 boys and 2 girls in the class.   **b:** ☐   **g:** ☐

   c) There are 9 girls and 20 children in the class.   **b:** ☐   **g:** ☐

   d) The ratio of boys to girls is 5:9 in the class.   **b:** ☐   **g:** ☐

   e) The ratio of girls to boys is 7:8 in the class.   **b:** ☐   **g:** ☐

   f) The ratio of boys to girls is 10:11 in the class.   **b:** ☐   **g:** ☐

4. From the information given, determine the number of girls and boys in each class.

   a) There are 20 children in a class. $\frac{2}{5}$ are boys.   b) There are 42 children. $\frac{3}{7}$ are girls.

   c) There are 15 children.   d) There are 24 children.
       The ratio of girls to boys is 3:2.   The ratio of girls to boys is 3:5.

NS6-108: Fractions, Ratios and Percents (continued)

5. Find the number of boys and girls in each classroom.

   a) In classroom A, there are 25 children: 60% are girls.

   b) In classroom B, there are 28 children. The ratio of boys to girls is 3 : 4.

   c) In classroom C, there are 30 children. The ratio of boys to girls is 1 : 2.

6. For each question below, say which classroom has more girls.

   a) In classroom A, there are 40 children. 60% are girls.

   In classroom B, there are 36 children. The ratio of boys to girls is 5 : 4.

   b) In classroom A, there are 28 children. The ratio of boys to girls is 5 : 2.

   In classroom B, there are 30 children. $\frac{3}{5}$ of the children are boys.

7. In the word "Whitehorse"…

   a) … what is the ratio of vowels to consonants?

   b) … what fraction of the letters are vowels?

   c) … what percent of the letters are consonants?

8. Look at the circle graph. Estimate the fraction of pets of each type owned according to the graph and complete the chart.

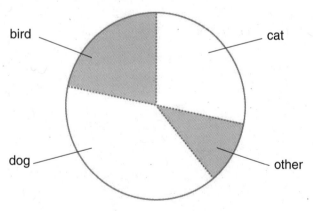

Pets		
Type	Fraction out of 100	Percent
dog		
cat		
bird		
other		

9. Write the amounts in order from least to greatest: $\frac{1}{20}$, 20%, 0.2. Show your work.

10. Kevin has 360 hockey cards.

    30% are Toronto Maple Leaf cards, and $\frac{1}{2}$ are Montreal Canadien cards. The rest are Vancouver Canuck cards.

    How many of each type of card does he have?

11. What percent of a metre stick is 37 cm? Explain.

NS6-109: 2-Digit Division

1. To divide a 3-digit dividend by a 2-digit divisor, start by estimating how many times the divisor goes into the dividend as shown below.

Step 1: Round the DIVISOR to the nearest ten and enter that number in the oval provided.

Step 2: Count by the leading digit of the rounded divisor to see how many times it goes into the DIVIDEND. Write your answer in the square.

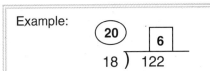

Example:

Step 1: Round 18 → 20

Step 2: Find out how many times 20 goes into 122, by skip counting or checking how many times 2 goes into 12 (= 6).

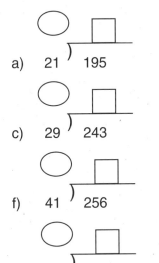

a) 21) 195

b) 19) 142

c) 29) 243

d) 42) 353

e) 48) 265

f) 41) 256

g) 49) 378

h) 32) 268

i) 62) 274

j) 29) 196

k) 28) 195

2. Next, multiply the divisor by the quotient.

Step 3: Multiply the DIVISOR by the quotient.

Step 4: Write the product underneath the DIVIDEND.

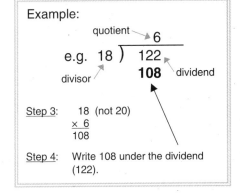

Example:

quotient → 6

e.g. 18) 122

divisor — dividend

108

Step 3: 18 (not 20)
× 6
108

Step 4: Write 108 under the dividend (122).

a) 6
41) 256

b) 6
28) 195

c) 7
19) 142

d) 9
21) 195

e) 4
62) 274

f) 7
49) 378

g) 6
29) 196

h) 8
29) 243

i) 8
42) 353

j) 8
32) 268

k) 5
48) 265

Number Sense 2

3. Complete Step 5 for <u>every</u> question before moving onto Step 6.

Step 5: Subtract.

Step 6: Write the remainder beside the QUOTIENT.

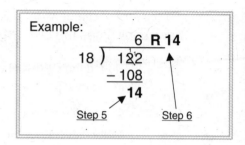

Example:

6 **R** 14
18) 122
−108
14

Step 5   Step 6

a) 42) 353
     − 336

b) 48) 265
     − 240

c) 49) 378
     − 343

d) 32) 268
     − 256

e) 62) 274
     − 248

f) 29) 196
     − 174

g) 28) 195
     − 168

h) 19) 142
     − 133

i) 41) 256
     − 246

j) 29) 243
     − 232

k) 21) 195
     − 189

4. Step 1: Round the DIVISOR to the nearest ten.

Step 2: Count by the leading digit of the rounded divisor to see how many times it goes into the DIVIDEND.

Step 3: Multiply the DIVISOR by the quotient.

Step 4: Write the product underneath the DIVIDEND.

Step 5: Subtract.

Step 6: Write the remainder beside the QUOTIENT.

a) 21) 156

b) 38) 249

c) 49) 358

d) 47) 326

e) 94) 419

f) 61) 559

g) 28) 192

h) 28) 219

i) 92) 293

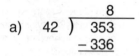

NS6-110: 2-Digit Division – Correcting Your Estimate

1. For each question, say whether the estimate was too high or too low.

Example:

$$\begin{array}{r} 7 \\ 23 \overline{)\ 156} \\ -\ 161 \end{array}$$

negative number!

ESTIMATE TOO HIGH!

$$\begin{array}{r} 6 \\ 16 \overline{)\ 123} \\ -\ 96 \end{array}$$

27   but 27 > 16

ESTIMATE TOO LOW!

a) $\begin{array}{r} 6 \\ 17 \overline{)\ 135} \\ -\ 102 \end{array}$

b) $\begin{array}{r} 6 \\ 23 \overline{)\ 129} \\ -\ 138 \end{array}$

c) $\begin{array}{r} 6 \\ 17 \overline{)\ 121} \\ -\ 102 \end{array}$

d) $\begin{array}{r} 4 \\ 26 \overline{)\ 149} \\ -\ 104 \end{array}$

e) $\begin{array}{r} 9 \\ 44 \overline{)\ 362} \\ -\ 396 \end{array}$

f) $\begin{array}{r} 6 \\ 24 \overline{)\ 126} \\ -\ 144 \end{array}$

2. In the space provided, correct the questions by calculating with the new estimate.

Example:

$$\begin{array}{r} 7 \\ 23 \overline{)\ 156} \\ -\ 161 \end{array}$$ 7 is too high so use 6 →

negative number!

ESTIMATE TOO HIGH!

$$\begin{array}{r} 6\ \textbf{R18} \\ 23 \overline{)\ 156} \\ -\ 138 \\ \textbf{18} \end{array}$$

a) $\begin{array}{r} 6 \\ 24 \overline{)\ 126} \\ -\ 144 \end{array}$ $24 \overline{)\ 126}$

negative number!
TOO HIGH!

b) $\begin{array}{r} 4 \\ 26 \overline{)\ 149} \\ -\ 104 \end{array}$ $26 \overline{)\ 149}$
45 45 > 26
TOO LOW!

c) $\begin{array}{r} 6 \\ 17 \overline{)\ 135} \\ -\ 102 \end{array}$ $17 \overline{)\ 135}$
33 33 > 17
TOO LOW!

d) $\begin{array}{r} 8 \\ 34 \overline{)\ 263} \\ -\ 272 \end{array}$ $34 \overline{)\ 263}$
negative number!
TOO HIGH!

e) $\begin{array}{r} 6 \\ 17 \overline{)\ 121} \\ -\ 102 \end{array}$ $17 \overline{)\ 121}$
19 19 > 17
TOO LOW!

f) $\begin{array}{r} 6 \\ 23 \overline{)\ 129} \\ -\ 138 \end{array}$ $23 \overline{)\ 129}$
negative number!
TOO HIGH!

g) $\begin{array}{r} 9 \\ 44 \overline{)\ 362} \\ -\ 396 \end{array}$ $44 \overline{)\ 362}$
negative number!
TOO HIGH!

3. a) $86 \overline{)\ 4677}$ b) $76 \overline{)\ 8460}$ c) $62 \overline{)\ 2486}$ d) $36 \overline{)\ 4175}$

4. A teacher divides 360 crackers among 24 students. How many crackers does each student get?

jump math
MULTIPLYING POTENTIAL

Number Sense 2

Answer the following questions in your notebook.

1. Meteorologists study the weather.

   The world's highest temperature in the shade was recorded in Libya in 1932.

   The temperature reached 58°C.

   a) How long ago was this temperature recorded?

   b) On an average summer day in Toronto, the temperature is 30°C.
      How much higher was the temperature recorded in Libya?

   c) The lowest temperature recorded (in Antarctica) was − 89°C.
      What is the difference between the lowest and the highest temperatures recorded?

2. The Olympic women's high jump gold medal was earned with a jump of 2.06 m.
   The silver jump was 2.02 m.

   a) Round both jumps to the nearest tenth.

   b) Make up two jumps which would round to the same number
      (when rounded to the tenths).

   c) Why are Olympic high jumps measured so precisely?

3. Doctors study the body. Here are some facts a doctor might know.

   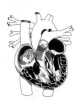

   a) FACT: "The heart pumps about 0.06 L of blood with each beat."
      About how many times would the heart need to beat to pump a litre of blood?

   b) FACT: "All of the blood passes through the heart in a minute."
      How many times would the blood pass through the heart in a day?

   c) FACT: "Bones make up about 15% of the weight of the body."
      How much would the bones of a 62 kg person weigh?

   d) FACT: "The brain is 85% water."
      What fraction of the brain is not water?

   e) FACT: "The most common type of blood is Type O blood.

      45% of people have Type O blood."

      About how many children in a class of 24 kids would have Type O blood?

NS6-112: Word Problems (Advanced)

Answer the following questions in your notebook.

1. 98% of Antarctica is covered in ice.

   What fraction of Antarctica is not covered in ice?

2. A ball is dropped from a height of 100 m. Each time it hits the ground, it bounces $\frac{3}{5}$ of the height it fell from. How high did it bounce…

   a) on the first bounce?
   b) on the second bounce?

3. The peel of a banana weighs $\frac{1}{8}$ of the total weight of a banana.

   If you buy 4 kg of bananas at $0.60 per kg …

   a) how much do you pay for the peel?
   b) how much do you pay for the part you eat?

4. The price of a soccer ball is $8.00. If the price rises by $0.25 each year,

   How much will the ball cost in 10 years?

5. Janice earned $28.35 on Monday.
   On Thursday, she spent $17.52 for a shirt.
   She now has $32.23.
   How much money did she have before she started work Monday?

6. Anthony's taxi service charges $2.50 for the first kilometre and $1.50 for each additional km.
   If Bob paid $17.50 in total, how many km did he travel in the taxi?

7. There are three apartment buildings in a block.

   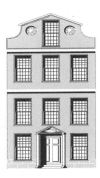

   - Apartment A has 50 suites.
   - Apartment B has 50% more suites than Apartment A.
   - Apartment C has twice as many suites as Apartment B.

   a) How many suites do Apartment B and Apartment C have?
   b) How many suites do the three apartments have altogether?

Answer the following questions in your notebook.

8. Six classes at Queen Victoria P.S. are going skating.

There are 24 students in each class.
The teachers ordered 4 buses, which each hold 30 students. Will there be enough room? Explain.

9. It took Cindy 20 minutes to finish her homework: she spent $\frac{2}{5}$ of the time on math and $\frac{2}{5}$ of the time on history.

   a) How many minutes did she spend on math and history?

   b) How many minutes did she spend on other subjects?

   c) What percent of the time did she spend on other subjects?

10. How many months old is a $1\frac{1}{2}$-year-old child?

11. Philip gave away 75% of his hockey cards.

   a) What fraction of his cards did he keep?

   b) Philip put his remaining cards in a scrapbook. Each page held 14 cards and he filled 23 pages.
   How many cards did he put in the book?

   c) How many cards did he have before he gave part of his collection away?

12. Find the mystery numbers.
   a) I am a number between 15 and 25.
      I am a multiple of 3 and 4.

   b) I am a number between 20 and 30.
      My tens digit is 1 less than my ones digit.

   c) Rounded to the nearest tens I am 60.
      I am an odd number.
      The difference in my digits is 2.

13. A pentagonal box has a perimeter of 3.85 m. How long is each side?

14. Tony bought a binder for $17.25 and a pen for $2.35. He paid 15% in taxes.
   How much change did he receive from $25.00?

15. Tom spent $500 on furniture: he spent $\frac{3}{10}$ of the money on a chair, $50.00 on a table and the rest on a sofa.
   What fraction and what percent of the $500.00 did he spend on each item?

ME6-8: Millimetres and Centimetres

If you look at a ruler with millimetre measurements, you can see that 1 cm is equal to 10 mm.

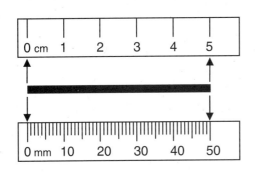

Measure the line in mm and cm.

The line is _____ cm long, or _____ mm long.

To convert a measurement from cm to mm,

we have to multiply the measurement by _____.

----

1. Your little finger is about 1 cm or 10 mm wide. Measure the objects below using your little finger.
   Then convert your measurement to mm.

   a)

   This pencil measures about _____ little fingers.

   So, the pencil is approximately _____ mm long.

   b)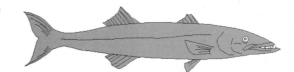

   This barracuda measures about _____ little fingers.

   So, the picture is approximately _____ mm long.

2. Find the distance between the two arrows on each ruler.

   a)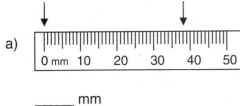

   _____ mm

   b)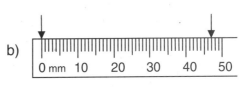

   _____ mm

3. Use a ruler to draw the following lines to the exact millimetre.

   a) Draw a line 27 mm long.   b) Draw a line 52 mm long.

4. a) Which is longer…

      i) Line A?   >————————<

         or Line B? <————————>

      ii) The height of the hat…
          or the brim of the hat?

          _____

   b) Measure the lengths in mm to check.

jump math
MULTIPLYING POTENTIAL.

Measurement 2

5. Estimate whether each line is <u>less</u> than 40 mm or <u>more</u> than 40 mm.
   Place a checkmark in the appropriate column.
   Then measure the actual length.

		Less than 40 mm	More than 40 mm
a)			
b)			
c)			

Actual Lengths:　a) _____ mm　　b) _____ mm　　c) _____ mm

6.

_____ cm

_____ cm

_____ mm　_____ cm

B

A

Measure the sides of the rectangle (in cm).

Then measure the distance between the two diagonal corners in cm and mm.

NOTE: Your answer in cm will be a decimal.

7. How many millimetres (mm) are there in one centimetre (cm)? _____

8. To change a measurement from centimetres (cm) into millimetres (mm), what should you <u>multiply</u> by?

_____

9. Fill in the missing numbers.

mm	cm
	13
	32

mm	cm
	8
	18

mm	cm
	213
	170

mm	cm
	9
	567

10. To change a measurement from mm to cm what should you <u>divide</u> by? _____

   a) 50 ÷ 10 = _____　　b) 80 ÷ 10 = _____　　c) 3200 ÷ 10 = _____　　d) 430 ÷ 10 = _____

   e) 460 mm = _____ cm　　f) 60 mm = _____ cm　　g) 580 mm = _____ cm

11. Convert.　　a) 4 cm = _____ mm　　b) 18 cm = _____ mm　　c) _____ cm = 130 mm

12. Circle the greater measurement in each pair of measurements below.
    First, convert one of the measurements so that both units are the same.

   a) 5 cm　　　70 mm　　b) 83 cm　　　910 mm　　c) 45 cm　　　53 mm

   d) 2 cm　　　12 mm　　e) 60 cm　　　6200 mm　　f) 72 cm　　　420 mm

ME6-8: Millimetres and Centimetres (continued)

13. Using your ruler, draw a second line so that the pair of lines are the given distance apart.

		Distance apart	
		in cm	in mm
a)		4	40
b)		3	_____
c)		_____	80
d)		7	_____

14. In the space provided, draw a line that is between 5 and 6 cm.

    How long is your line in mm? _____

15. Write a measurement in mm that is between …

    a) 7 and 8 cm: ____ mm   b) 12 and 13 cm: _____   c) 27 and 28 cm: _____

16. Write a measurement in a whole number of cm that is between …

    a) 67 mm and 75 mm: ___ cm   b) 27 mm and 39 mm: _____   c) 52 mm and 7 cm: _____

17. Draw a line that is a whole number of centimetres long and is between …

    a) 35 and 45 mm   b) 55 and 65 mm   c) 27 and 33 mm

18. Rebecca says 7 mm is longer than 3 cm because 7 is greater than 3. Is she right?

19. Carl has a set of sticks: some are 7 cm long and some are 4 cm long.

    Example: This picture (not drawn to scale) shows how
                he could line up the sticks to measure 19 cm: _7 cm_ _4 cm_ _4 cm_ _4 cm_

    Draw a sketch to show how Carl could measure each length by lining the sticks up end to end.

    a) 8 cm   b) 11 cm   c) 22 cm   d) 26 cm   e) 25 cm

20. Show how Carl could make these measurements using his sticks.
    HINT: you may need to subtract.

    a) 3 cm   b) 1 cm   c) 20 mm   d) 50 mm   e) 17 cm

    BONUS:
    f) Can you find two different solutions for each measurement?

ME6-9: Decimetres

A **decimetre** is a unit of length equal to 10 cm.

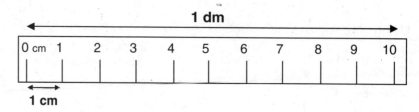

1 dm

| 0 cm | 1 | 2 | 3 | 4 | 5 | 6 | 7 | 8 | 9 | 10 |

1 cm

---

1. Place a checkmark in the correct column.

   HINT: You can use the picture at the top of the page to help you estimate.

	Less than 1 dm	More than 1 dm
My leg		
The length of an eraser		
My pencil		
The height of the classroom door		

2. 1 decimetre = _____ centimetres.

3. What fraction of a decimetre (dm) is a centimetre? _____

4. To change a measurement from dm to cm, what should you <u>multiply</u> by? _____

5. To change a measurement from cm to dm what should you <u>divide</u> by? _____

6. Find the numbers missing from the following charts.

cm	dm
120	12
	31
	42

cm	dm
80	
	620
300	

cm	dm
530	
	1
950	

7. In the space provided, draw a line that is between 1 and 2 decimetres long.

   a) How long is your line in cm? _____   b) How long is your line in mm? _____

8. Write a measurement in cm that is between …

   a) 3 and 4 dm _____   b) 6 and 7 dm _____   c) 9 and 10 dm _____

9. Write a measurement in dm that is between …

   a) 62 and 72 cm _____   b) 37 and 45 cm _____   c) 48 and 73 cm _____

10. How many dm are in 100 cm? _____

11. There are 10 mm in 1 cm. There are 10 cm in 1 dm. How many mm are in 1 dm? _____

ME6-10: Metres and Kilometres

A **metre** is a unit of measurement for <u>length</u> (or <u>height</u> or <u>thickness</u>) equal to 100 cm.

A metre stick is 100 cm long.

A **kilometre** is a unit of measurement for length equal to 1 000 metres.

Here are some measurements you can use for estimating in metres.

about **2** metres:

the height of
a (tall) adult

about **10** metres:

the length of
a school bus

about **100** metres:

the length of
a football field

1. Find (or think of) an object in your classroom or outside that is approximately …

   a) 2 metres long _____   b) 3 metres long _____

2. Fourteen basketball players can lie head to foot along a basketball court.

   What is the court's length in metres? _____

3. a) How many adults do you think could lie head to foot across your classroom? _____

   b) Approximately how wide is your classroom (in metres)? _____

4. a) About how many school buses high is your school? _____

   b) About how high is your school (in metres)? _____

5. A small city block is about 100 m long.

   Name a place you can walk to from your school: _____

   Approximately how many metres away from the school is the place you named? _____

6. The number line represents 1 km. Mark the following distances on the line:

   A 200 m   **B** 50 m   **C** 550 m   **D** 825 m   **E** 110 m

   0 km |___|___|___|___|___|___|___|___|___|___| 1 km

7. About how many football fields long is a kilometre?

8. You can travel 1 km if you walk for 15 minutes at a regular speed.
   Name a place that is about 1 km from your school.

jump math
MULTIPLYING POTENTIAL.

Measurement 2

ME6-11: Comparing Units

1. Finish the table by following the pattern.

m	1	2	3	4	5	6
dm	10	20				
cm	100	200				
mm	1000	2000				

2. To convert a measurement from metres to centimetres, you multiply by _____.

3. To convert a measurement from metres to millimetres, you multiply by _____.

4. Convert the following measurements.

m	cm
8	
70	

m	mm
5	
17	

cm	mm
4	
121	

dm	cm
32	
5	

5. Convert the measurement given in cm to a measurement using multiple units.

   a) 423 cm = ___ m _____ cm   b) 514 cm = ___ m _____ cm   c) 627 cm = ___ m _____ cm

   d) 673 cm = ___ m _____ cm   e) 381 cm = ___ m _____ cm   f) 203 cm = ___ m _____ cm

6. Convert the following multiple units of measurements to a single unit.

   a) 2 m 83 cm = _____ cm   b) 3 m 65 cm = _____ cm   c) 4 m 85 cm = _____ cm

   d) 9 m 47 cm = _____ cm   e) 7 m 4 cm = _____ cm   f) 6 m 40 cm = _____ cm

7. Change the following measurements to multiple units then to decimal notation.

   a) 546 cm = __5_ m __46__ cm = __5.46__ m   b) 217 cm = ____ m _____ cm = _____ m

   c) 783 cm = ____ m _____ cm = _____ m   d) 608 cm = ____ m _____ cm = _____ m

   e) 72 cm = ____ m _____ cm = _____ m   f) 7 cm = ____ m _____ cm = _____ m

8. Why do we use the same decimal notation for dollars and cents and for metres and centimetres?

9. Michelle says that to change 6 m 80 cm to centimetres, you multiply the 6 by 100 and then add 80.
   Is Michelle correct? Why does Michelle multiply by 100?

ME6-12: Changing Units

1. Measure the line below in mm, cm, and dm:

_____

_____ mm   _____ cm   _____ dm

a) Which of the units (mm, cm, or dm) is: largest? _____ smallest? _____

b) Which unit did you need more of to measure the line, the <u>larger</u> unit or the <u>smaller</u> unit?

_____

c) To change a measurement from a **larger** to a **smaller** unit, do you need ...

(i) more of the smaller units, or   **(ii) fewer** of the smaller units?

_____

2. Fill in the missing numbers.

a) 1 cm = _____ mm   b) 1 dm = _____ cm

c) 1 dm = _____ mm   d) 1 m = _____ dm

e) 1 m = _____ cm   f) 1 m = _____ mm

Units **decrease** in size going **down** the stairway:	m
	⌐ dm
	  ⌐ cm
	    ⌐ mm

- 1 step down   = 10 × smaller
- 2 steps down = 100 × smaller
- 3 steps down = 1 000 × smaller

3. Change the measurements below by following the steps.

a) Change 3.5 cm to mm

   i) The new units are __10__ times ___smaller___

   ii) So I need __10__ times ___more___ units

   iii) So I ___multiply___ by __10__

   3.5 cm = __35__ mm

b) Change 7.2 cm to mm

   i) The new units are ____ times _____

   ii) So I need ____ times _____ units

   iii) So I _____ by ____

   7.2 cm = ____ mm

c) Change 2.6 dm to cm

   i) The new units are ____ times _____

   ii) So I need _____ times _____ units

   iii) So I _____ by _____

   2.6 dm = ____ cm

d) Change 7.53 cm to mm

   i) The new units are ____ times _____

   ii) So I need ____ times _____ units

   iii) So I _____ by _____

   7.53 cm = ____ mm

Measurement 2

REMEMBER: There are 1 000 g in a kg and 1 000 mg in a g.

e) Change 3.4 mm to cm

  i)   The new units are _____ times_____

  ii) So I need _____ times _____ units

  iii) So I _____ by _____

      3.4 mm = _____ cm

f) Change 8.53 kg to g

  i)   The new units are _____ times _____

  ii) So I need _____ times _____ units

  iii) So I _____ by _____

      8.53 kg = _____ g

g) Change 5.2 g to mg

  i)   The new units are _____ times _____

  ii) So I need _____ times _____ units

  iii) So I _____ by _____

      5.2 g = _____ mg

h) Change 2.14 g to kg

  i)   The new units are _____ times _____

  ii) So I need _____ times _____ units

  iii) So I _____ by _____

      2.14 g = _____ kg

4. Change the units by following the steps in Question 3 mentally.

  a) 4 m = _____ dm

  b) 1.3 dm = _____ mm

  c) 20 cm = _____ mm

5. A decimetre of ribbon costs 5¢.
    How much will 90 cm cost?

6. Emily's books weigh 2.1 kg, 350 g, and 1253 g.
    Her backpack can hold 4 kg.
    Can she carry all three books?

7. The width of a rectangle is 57 cm and its length is 65 cm.
    Is the perimeter of the rectangle greater or less than 2.4 m?

8. How is the relation between kilograms and grams similar to the relation between kilometres and metres?

9. How is the relation between milligrams and grams similar to the relation between millimetres and metres?

ME6-13: Appropriate Units of Length

1. Match the word with the symbol.
    Then match the object with the appropriate unit of measurement.

    a)

mm	kilometre
cm	centimetre
m	millimetre
km	metre

length of a bee's antenna
width of a swimming pool
distance of a marathon
diameter of a drum

    b)

km	metre
cm	millimetre
m	kilometre
mm	centimetre

length of a ruler
thickness of a nail
diameter of the moon
length of a soccer field

2. Circle the unit of measurement that makes the statement correct.

    a) A very tall adult is about 2 **dm** / **m** high.

    b) The width of your hand is close to 1 **dm** / **cm**

    c) The Calgary Tower is 191 **cm** / **m** high.

3. Nicholas measured some objects, but forgot to include the units. Add the appropriate unit.

    a) bed: 180 _____   b) car: 2 _____   c) hat: 25 _____   d) toothbrush: 16 _____   e) driveway: 11 _____

4. Choose which unit (km, m, or cm) belongs to complete each sentence.

    a) Canada's entire coastline is 202 080 _____ long.

    b) Mount Logan, in the Yukon Territory, is 5 959 _____ high.

    c) Water going over the Della Falls in British Columbia drops 440 _____.

    d) A teacher's desk is about 200 _____ long.

    e) A fast walker can walk 1 _____ in 10 minutes.

    f) A great white shark can grow up to 4 _____.

    g) A postcard is about 15 _____ long.

5. Most provinces in Canada have an official tree and an official bird.
    Change each measurement below to the smallest unit used.
    Then order the trees from tallest to shortest.

Tree	Height	In the Smallest Units
White Birch (Saskatchewan)	2 m	
Lodgepole Pine (Alberta)	3 050 cm	
Western Red Cedar (British Columbia)	59 m	
Red Oak (Prince Edward Island)	24 m	

1. _____
2. _____
3. _____
4. _____

jump math
MULTIPLYING POTENTIAL.

Measurement 2

6. Order the official birds from <u>longest</u> to <u>shortest</u>.

Bird	Length	In the Smallest Units
Atlantic Puffin (Newfoundland & Labrador)	34.5 cm	
Great Horned Owl (Alberta)	63.5 cm	
Snowy Owl (Quebec)	66 cm	
Great Gray Owl (Manitoba)	0.55 m	

1. _____

2. _____

3. _____

4. _____

7. Mark each measurement on the number line with an 'X'. The first one is done for you.

A 12 mm   B 35 mm   C 2.0 cm   D 49 mm   E 9.9 cm   F 5.7 cm

0 dm |_____| 1 dm

G 3 cm   H 5 cm   I 25 mm   J 9 cm   K 4.5 cm   L 8.2 cm

0 km |_____| 1 km

M 200 m   N 500 m   O 700 m   P 350 m   Q 850 m   R 630 m   S 90 m

8. Fill in the numbers in the correct places below (select from the box).

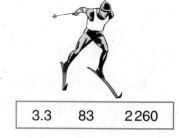

a) The 1988 Winter Olympics skiing competitions took place at
   Nakiska, which is _____ **km** from Calgary.

   The highest point at Nakiska is _____ **m** above sea level.

   The longest run down the mountain is _____ **km.**

3.3	83	2 260

b) The Red River is _____ **km** long.

   In 1997 it flooded and rose _____ **m.**

   Winnipeg was protected by a _____ **km** floodway that was
   built around the city.

7.5	47	877

9. Name an object in your classroom that has …

   a) a thickness of about 20 mm: _____   b) a height of about 2 m: _____

ME6-14: Speed

Speed is the rate of motion or the distance covered in a certain time.
A standard measure of speed is kilometres per hour (km/h).
If you travel 40 km in 1 hour, then your speed is 40 km/h.

Average speed is found by dividing the total distance travelled by the amount of time spent travelling.

   Example: Anu cycled 75 km in 5 hours. Her average speed was 75 ÷ 5 = 15 km/h.

----

1. Find the average speed for each set of distances and times. (Try to find the answer mentally.)

	Distance	Time	Average Speed
a)	200 km	2 hours	
c)	75 km	3 hours	

	Distance	Time	Average Speed
b)	480 km	6 hours	
d)	1600 km	10 hours	

2. Fill in the charts below.
   HINT: Average Speed × Time = Distance

	Average Speed	Time	Distance
a)	25 km/h	1 hour	
c)	70 km/h	3 hours	

	Average Speed	Time	Distance
b)	120 km/h	3 hours	
d)	30 km/h	7 hours	

3. During a sneeze, air moves at 167 km/h.
   During a hurricane, air moves at 117.5 km/h.
   How much faster is a sneeze than a hurricane?

4. Clare can cycle at a speed of 23 km/h.
   Erin can cycle at a speed of 17 km/h.
   How much further can Clare cycle in 3 hours than Erin?

5. a) A truck travels 40 kilometres in half an hour.
      What is its average speed in km/h?

   b) A car travels 30 kilometres in 15 minutes.
      What is its average speed in km/h?

6. Jinny walked 5.25 kilometres in 5 hours and Paula walked 6.23 kilometres in 7 hours.
   Whose average speed was greater?

7. Helen walked 4 km in an hour and then cycled 16 km in the second hour.

   a) How far did she travel?   b) What was her average speed?

jump math
MULTIPLYING POTENTIAL.

Measurement 2

Answer the following questions in your notebook.

The diagrams show the tallest buildings in each province.
NOTE: The CN Tower in Toronto has not been included as it is more of a "structure" than a "building".

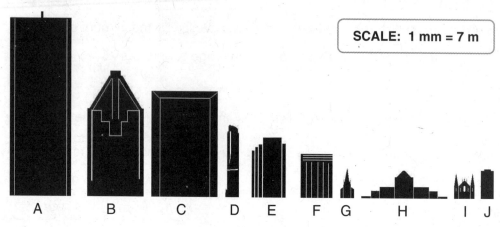

SCALE: 1 mm = 7 m

1. Choose two buildings.

   Measure each diagram, and calculate the height of each building using the scale.
   Find the difference in height between the buildings.

	Building		Building
A	First Canadian Place, ON	B	Le 1000 Rue de la Gauchetière, QC
C	Petro Canada Centre, AB	D	Wall Centre, BC
E	Toronto Dominion Centre, MB	F	Fenwick Tower, NS
G	Cathedral of the Immaculate Conception, NB	H	Confederation Building, NL
I	Saint Dunstan's Roman Catholic Cathedral, PE	J	CN Tower, SK

2. The CN Tower in Saskatchewan has the same name as the CN Tower in Toronto.
   The CN Tower in Toronto is 553 m tall.

   a) About how much taller is the Toronto tower than the Saskatchewan tower?

   b) About how many of the Saskatchewan towers stacked on top of each other would be the same height as the Toronto tower?

3. The tallest residential building is Eureka Tower in Melbourne, Australia.
   It is about one and a half times taller than Petro Canada Centre in Calgary.
   If you had to draw it to the scale 1 mm = 7 m, what should the height of your drawing be?

4. a) Wall Centre in Vancouver has 48 floors. First Canadian Place in Toronto has 72 floors.
      Can you tell which of these two buildings have a greater amount of height per floor without long division? Explain.

   b) Use long division to find the height per floor of these two buildings.

ME6-16: The Welland Canal

The Welland Canal connects Lake Ontario and Lake Erie, allowing large ships to pass between the two lakes. Ships are raised and lowered through a system of locks.

On the map, 1 mm represents about 200 m.

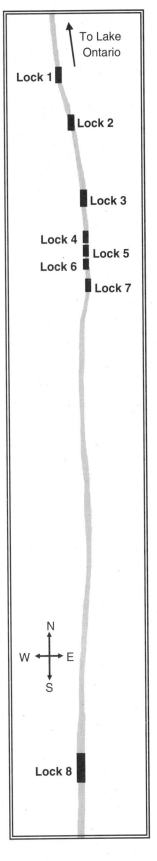

1. Using your ruler, measure the distance between the given locks. Then find the actual distance by using the scale.

   Lock 1 and Lock 2

   Distance on map: _____   Actual distance: _____

   Lock 2 and Lock 3

   Distance on map: _____   Actual distance: _____

   Lock 7 and Lock 8

   Distance on map: _____   Actual distance: _____

2. Could a boat travelling at 10 km/h go from Lock 7 to Lock 8 in two hours?

3. The total length of the canal is approximately 44 km.
   About how long would it take you to travel the canal at 10 km/h, stopping for a half hour at each lock?

4. The Garden City Skyway runs over the canal about 1.8 km south of Lock 2. Draw a line where you think the Skyway should go.

5. The largest canal in the world is the Grand Canal in China at 1795 km. About how many times would you travel up and down the Welland Canal to travel the same distance?

1. Change the amount given from dollars and cents to cents.

   a) 5 dollars 28 cents = _____528 cents_____

   b) 7 dollars 14 cents = _____

   c) 10 dollars 3 cents = _____

   d) 4 dollars 8 cents = _____

2. Change the measurement from meters and centimetres to centimetres.

   a) 6 m 2 cm = ____602 cm____

   b) 4 m 9 cm = _____

   c) 6 m 13 cm = _____

   d) 11 m 53 cm = _____

   e) 14 m 20 cm = _____

   f) 3 m 1 cm = _____

3. Change the measurement from kilometres and meters to meters.

   a) 9 km 2 m = ____9002 m____

   b) 4 km 73 m = _____

   c) 5 km 10 m = _____

   d) 6 km 2 m = _____

   e) 13 km 241 m = _____

   f) 20 km 2 m = _____

4. Change from hours and minutes to minutes. Remember, there are 60 minutes in an hour.

   a) 3 h 3 min = ____183 min____

   b) 2 h 4 min = _____

   c) 3 h 11 min = _____

   d) 4 h 10 min = _____

   e) 3 h 52 min = _____

   f) 5 h 26 min = _____

5. Change each amount to a decimal in the larger unit.

   a) $7 and 3¢ = ____$7.03____

   b) $16 and 4¢ = _____

   c) $27 and 3¢ = _____

   d) 8 m 6 cm = _____

   e) 9 m 25 cm = _____

   f) 3 cm 1 mm = _____

   g) 9 dm 4 cm = _____

   h) 8 L 7 mL = _____

   i) 9 kg 25 g = _____

6. Giant kelp can grow up to 6 dm a day.
   How many metres can it grow in 4 weeks?

ME6-18: Exploring Perimeter

1. Each edge is 1 cm long. Write the total length of each side in cm as shown in the first figure. Then write an addition statement and find the perimeter.

a)

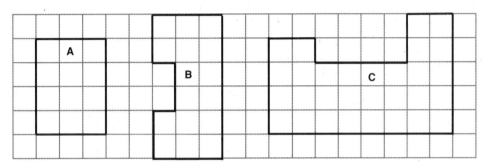

b)

Perimeter: _____

Perimeter: _____

2. Each edge is 1 unit long. Write the length of each side beside the figure (don't miss any edges!). Then use the side lengths to find the perimeter.

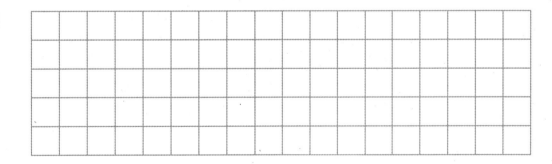

3. Draw your own figure and find the perimeter.

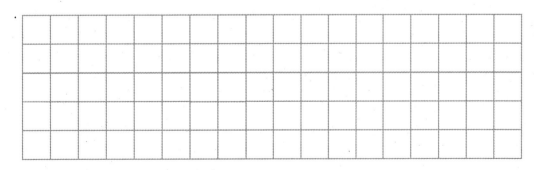

4. Draw two different shapes with a perimeter of 12 units.

5. On grid paper, draw your own figures and find their perimeters. Try making letters or other shapes!

ME6-19: Measuring Perimeter

1. Measure the perimeter of each figure in cm using a ruler.

   a)

   Perimeter: _____

   b)

   Perimeter: _____

   c)

   Perimeter: _____

2. Find the perimeter of each shape. Be sure to include the units in your answer.

   a)
   7 m
   5 m   A

   Perimeter: _____

   b)
   3 cm
   2 cm
   6 cm   B   5 cm
   4 cm
   8 cm

   Perimeter: _____

   c)
   2 km   C   2 km
   2 km

   Perimeter: _____

   d)
   5 cm
   D   10 cm

   Perimeter: _____

   e) Write the letters of the shapes in order from <u>greatest</u> perimeter to <u>least</u> perimeter.
      HINT: Make sure you look at the units!

      _____

3. Your little finger is about 1 cm wide. Estimate, then measure, the perimeter of each shape in cm.

   a)

   Estimated Perimeter: _____

   Actual Perimeter: _____

   b)

   Estimated Perimeter: _____

   Actual Perimeter: _____

4. Show all the ways you can make a rectangle using …

   a) 10 squares   b) 12 squares   c) Can you make a rectangle with 7 squares?

   d) Which of the rectangles in b) has the greatest perimeter? What is the perimeter?

5. Draw three different figures with perimeter 10.
   NOTE: The figures don't have to be rectangles.

6. A rectangle has perimeter 1 m. Each longer side is 36 cm. How long is each shorter side?

7. A rectangle is twice as long as it is wide.
   What is the ratio of the width to the perimeter of the rectangle?

1. Mark makes a sequence of figures with toothpicks.

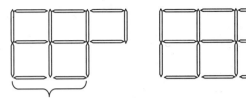

base

a) Complete the chart.

b) Complete the rule that tells how to make the OUTPUT
    numbers from the INPUT numbers.

    Multiply the INPUT by _____ and add _____.

c) Use the rule to predict the perimeter of
    a figure with a base of 10 toothpicks: _____

INPUT Number of toothpicks in base	OUTPUT Perimeter
2	10

2. Add one square to the figure so that the perimeter of the new figure is 12 units.
    NOTE: Assume all edges are 1 unit.

    a)   b) 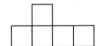   c)

    Original Perimeter = ____ units   Original Perimeter = ____ units   Original Perimeter = ____ units

    New Perimeter = 12 units   New Perimeter = 12 units   New Perimeter = 12 units

3. Find all rectangles with the given perimeter (with lengths and widths that are whole numbers).

Width	Length

Perimeter = 6 units

Width	Length

Perimeter = 12 units

Width	Length

Perimeter = 16 units

Width	Length

Perimeter = 18 units

4. Repeat steps a) to c) of Question 1 for the following patterns.

    a)    b)

5. Emma says the formula 2 × (length + width) gives the perimeter of a rectangle.
    Is she correct?

ME6-21: Circles and Irregular Polygons

1. The horizontal and vertical distance between adjacent pairs of dots is 1 cm. The diagonal distance is about 1.4 cm.

   1.4 cm

   Find the approximate perimeter of each figure by counting diagonal sides as 1.4 cm.
   HINT: How can multiplication help you sum the sides of length 1.4 cm?

   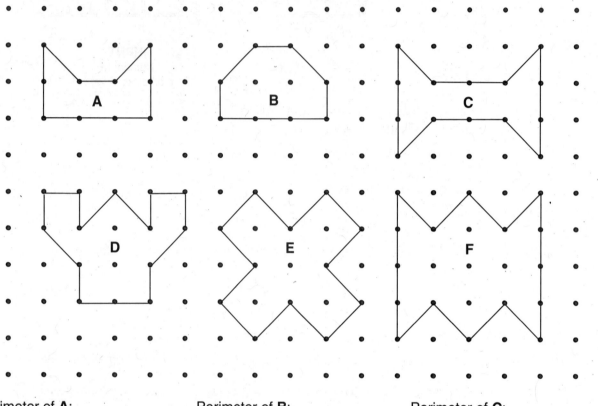

   Perimeter of **A**: _____   Perimeter of **B**: _____   Perimeter of **C**: _____

   Perimeter of **D**: _____   Perimeter of **E**: _____   Perimeter of **F**: _____

2. The distance around the outside of a circle is called the **circumference**.

   a) Measure the circumference of each circle to the nearest **cm** using a strip of rolled up paper and a ruler. Record the width and circumference in the chart.

Width	Circumference

   top view

   mark the distance around the circle on the strip

   side view

   unroll your strip and measure the distance with your ruler

   b) About how many times greater than the width is the circumference? _____

ME6-22: Area in Square Centimetres

Shapes that are flat are called **two-dimensional** (2-D) shapes.

The **area** of a 2-dimensional shape is the amount of space it takes up.

A **square centimetre** is a unit for measuring area. A square with sides of 1 cm has an area of one square centimetre. The short form for a square centimetre is cm².

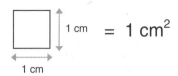

1. Find the area of these figures in square centimetres.

a)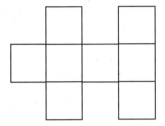

Area = _____ cm²

b)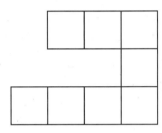

Area = _____ cm²

c)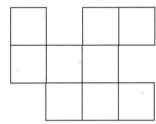

Area = _____ cm²

2. Using a ruler, draw lines to divide each rectangle into square centimetres.

a)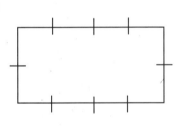

Area = _____ cm²

b)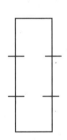

Area = _____ cm²

c)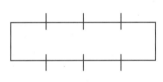

Area = _____ cm²

3. How can you find the area (in square units) of each of the given shapes?

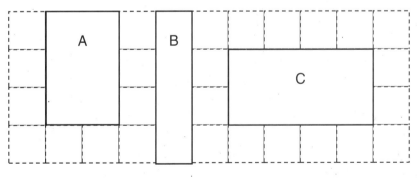

Area of **A** = _____   Area of **B** = _____   Area of **C** = _____

4. Draw three different shapes that have an area of 10 cm² (the shapes don't have to be rectangles).

5. Draw several shapes and find their area and perimeter.

6. Draw a rectangle with an area of 12 cm² and perimeter of 14 cm.

Measurement 2

ME6-23: Area of Rectangles

1. Write a multiplication statement for each array.

a) 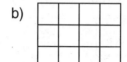   b)   c)   d)

_____   _____   _____   _____

2. Draw a dot in each box. Then write a multiplication statement that tells you the number of boxes in the rectangle.

a) 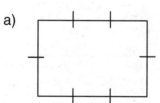   b)   c)   d)

___3 × 7 = 21___   _____   _____   _____

3. Write the number of boxes along the width and length of each rectangle. Then write a multiplication statement for the area of the rectangle (in square units).

a)   Width =   b)   Width =   c)   Width =
___   ___   ___

Length = ____   Length = ____   Length = ____

_____   _____   _____

4. Using a ruler, draw lines to divide each rectangle into squares. Write a multiplication statement for the area of the boxes in cm^2.
NOTE: You will have to mark the last row of boxes yourself using a ruler.

a)   b)   c)

d)   e)

5. If you know the length and width of a rectangle, how can you find its area?

_____

ME6-24: Exploring Area

1. Measure the length and width of the figures then find the area.

    a)

    b)

    c)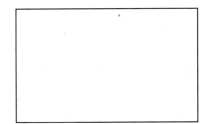

    _____   _____   _____

2. Find the area of a rectangle with the following dimensions:

    a) width: 6 m   length: 7 m   b) width: 3 m   length: 7 m   c) width: 4 cm   length: 8 cm

    _____   _____   _____

3. a) Calculate the area of each rectangle. Be sure to include the units.

    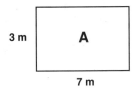   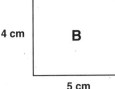

    Area: _____   Area: _____   Area: _____   Area: _____

    b) By letter, create an ordered list of the rectangles from <u>greatest</u> to <u>least</u> area: _____

4. A rectangle has an area of 18 cm^2 and a length of 6 cm. How can you find its width?

    _____

5. A rectangle has an area of 24 cm^2 and a width 8 cm. What is its length? _____

6. A square has an area of 25 cm^2. What is its width? _____

7.

    a) Write the lengths of each side on the figure.

    b) Divide the figure into two boxes.

    c) Calculate the area by finding the area of the two boxes.

        Area Box 1: _____   Area of Box 2: _____

        TOTAL Area: _____

8. Using grid paper or a geoboard, create two different rectangles with an area of 12 square units.

jump math
MULTIPLYING POTENTIAL.

Measurement 2

ME6-25: Comparing Area and Perimeter

1. For each shape below, calculate the perimeter and area of each shape, and write your answers in the chart below. The first one has been done for you.

   NOTE: The edge of each grid square represents 1 cm.

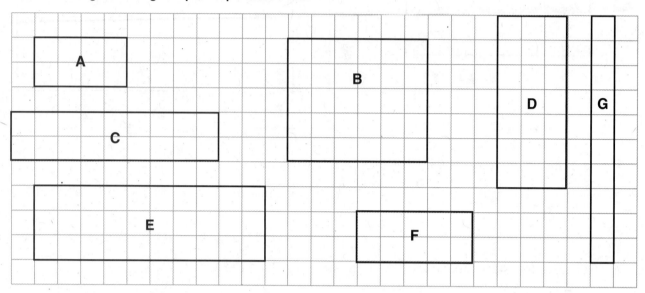

Shape	Perimeter	Area
A	2 + 4 + 4 + 2 = 12 cm	$2 \times 4 = 8$ cm^2
B		
C		
D		
E		
F		
G		

2. Shape C has a greater perimeter than shape D. Does it also have greater area? _____

3. Name two other shapes where one has a greater perimeter and the other, a greater area:

   _____

4. Write the shapes in order from greatest to least perimeter: _____

5. Write the shapes in order from greatest to least area: _____

6. Are the orders in Questions 4 and 5 the same? _____

7. What is the difference between <u>perimeter</u> and <u>area</u>? _____

   _____

   _____

ME6-26: Area and Perimeter

1. Measure the length and width of each rectangle, and then record your answers in the chart below.

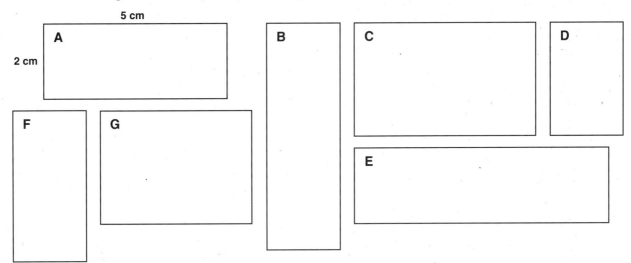

Rectangle	Estimated Perimeter	Estimated Area	Length	Width	Actual Perimeter	Actual Area
A	cm	cm²	cm	cm	cm	cm²
B						
C						
D						
E						
F						
G						

2. Measure the perimeter and find the area for each rectangle below with a ruler.

a)

Perimeter = _____ cm

Area = _____ cm²

b)

Perimeter = _____ cm

Area = _____ cm²

c)

Perimeter = _____ cm

Area = _____ cm²

3. Find the area of the rectangle using the clues.

a) Width = 2 cm; Perimeter = 10 cm; Area = ?   b) Width = 4 cm; Perimeter = 18 cm; Area = ?

4. Draw a square on grid paper with the given perimeter. Then find the area of the square.

a) Perimeter = 12 cm; Area = ?   b) Perimeter = 20 cm; Area = ?

5. On grid paper, draw a rectangle with …

a) an area of 10 square units and a perimeter of 14 units.

b) an area of 8 square units and a perimeter of 12 units.

ME6-27: Area of Polygons and Irregular Shapes

1. Two half squares  cover the same area as a whole square 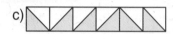.

Count each <u>pair</u> of half squares as a whole square to find the area shaded.

a)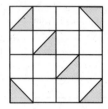

= _____ whole squares

b)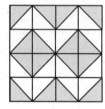

= _____ whole squares

c)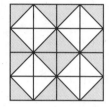

= _____ whole squares

d)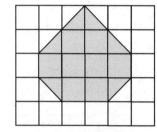

= _____ whole squares

e)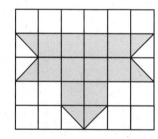

= _____ whole squares

f)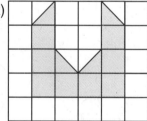

= _____ whole squares

g)

= _____ whole squares

h)

= _____ whole squares

i)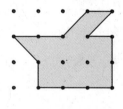

= _____ whole squares

j)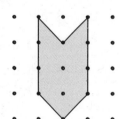

= _____ whole squares

k)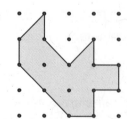

= _____ whole squares

2. Estimate then find the area of each figure in square units.
   HINT: Draw lines to show all the half squares.

a)

b)

c)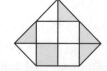

3. For each picture say whether the shaded area is <u>more</u> than, <u>less</u> than or <u>equal</u> to the unshaded area. Explain how you know.

a)

b)

c)

jump math
MULTIPLYING POTENTIAL

Measurement 2

4.

   a) What fraction of the rectangle is the shaded part? _____

   b) What is the area of the rectangle in square units? _____

   c) What is the area of the shaded part? _____

5. Find the shaded area in square units.

  a)    b) 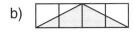   c)    d)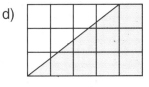

    _____   _____   _____   _____

6. Draw a line to divide each shape into two triangles or a triangle and a rectangle.
   Then calculate the area of each shape.

  a) 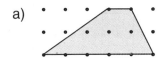   b) 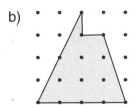   c) 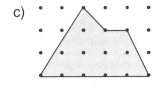   d)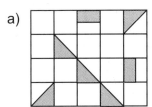

    _____   _____   _____   _____

7. Calculate the area of each shape.

  a) 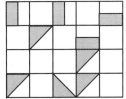   b) 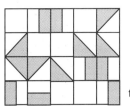   c)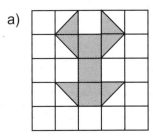

8. Each of the shaded shapes below represents ½ a square. How many total squares do they add up to?
   REMEMBER: Two ½ squares = 1 full square

  a)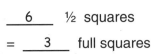

    _____ half squares

    _____ total squares

  b)

    _____ half squares

    _____ total squares

  c)

    _____ half squares

    _____ total squares

9. Fill in the blanks to find the total area. The first one has been done for you.

  a)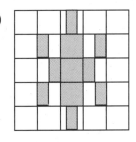

    __3__ full squares

    __6__ ½ squares

    = __3__ full squares

    Area = 3 + 3 = 6

  b)

    _____ full squares

    _____ ½ squares

    = _____ full squares

    Area =

10. Estimate the areas of the shaded figures below as follows.

    • Put a check mark in each <u>half</u> square: 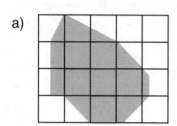, etc.

    • Put an 'X' in every <u>full</u> square **and** in every square with <u>more than half</u> shaded: 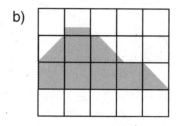, etc.

    • Count all squares with an 'X' as 1. Count 2 half squares (marked with a check) as 1.

    • Do not count squares where <u>less than half</u> is shaded: ▢, ◹, etc.

a)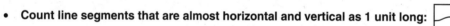

b)

_____ half squares (= _____ full squares)

+ _____ full squares

= _____ total squares

_____ half squares (= _____ full squares)

+ _____ full squares

= _____ total squares

11. Estimate the area (in square units) and perimeter of the shapes below.
    HINT: For estimating perimeter ...

    • **Count line segments that are almost horizontal and vertical as 1 unit long:**

    • **Count line segments that are almost diagonal as $1\frac{1}{2}$ (or 1.5):**

    • **Count line segments that are close to half as $\frac{1}{2}$:**

a)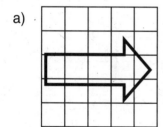

b)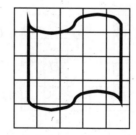

c)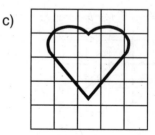

Approximate
Area: _____

Approximate
Perimeter: _____

Approximate
Area: _____

Approximate
Perimeter: _____

Approximate
Area: _____

Approximate
Perimeter: _____

ME6-28: Problems and Puzzles

1.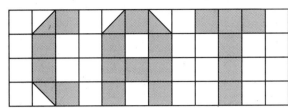

a) Find the area of the shaded pattern block word.

b) There are 48 squares in the grid. How can you use your answer above to find the number of <u>unshaded</u> squares (without counting them)?

2. Ed bakes a rectangular birthday cake for his Dad.

The cake will be cut into twenty-four 5 × 5 cm pieces.

a) What is the area of the cake?

b) The width of the cake is 20 cm. What is its length?

c) Ed puts blackberries on the perimeter of the cake, 2 blackberries on each 5 cm. How many berries does he need?

d) Blackberries are sold in packs of 20 berries. Each pack costs $2.99. If Ed pays for the blackberries with a 10-dollar bill, how much change does he get?

3. A rectangle's length and width are both whole numbers, where the length is greater than the width. Find possible sizes for the rectangle, given the areas below.

Area = 8 cm²	
Length	Width

Area = 14 cm²	
Length	Width

Area = 18 cm²	
Length	Width

4. Name something you would measure in …

a) square metres _____

b) square kilometres _____

_____

_____

BONUS

5. Find the area of the shaded part. Then, say what fraction of the grid is shaded.
   HINT: How can you use the area of the unshaded part and the area of the grid?

a)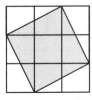

Area:

_____

Fraction:

_____

b)

Area:

_____

Fraction:

_____

c)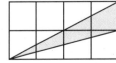

Area:

_____

Fraction:

_____

 jump math
MULTIPLYING POTENTIAL.

Measurement 2

ME6-29: Area and Perimeter (Advanced)

1. There are 100 cm² in 1 dm².

    a) How many cm² are in 1 m²? _____

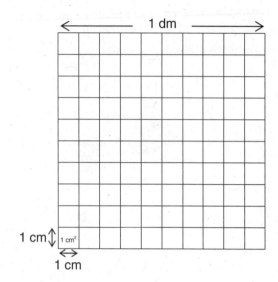 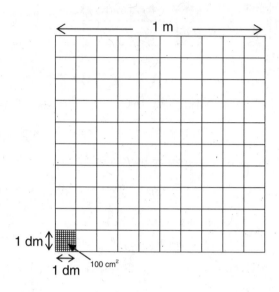

    b) Change 1.34 m² to cm².

    The new units are _____ times _____. So I need _____ times _____ units.
    larger/smaller   _more/fewer_

    So I _____ by _____. So 1.34 m² = _____ cm².
    multiply/divide

    c) 14.65 m² = _____ cm²   d) 0.01 m² = _____ cm²

    e) 0.376 m² = _____ cm²   f) 7.2 m² = _____ cm²

2. (i) Find the ratios of the perimeter of each square to the length of its side.

    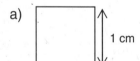 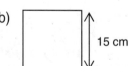  

    (ii) What do you notice?

3. In each rectangle the long side is twice as long as the short side. Find the ratio of the perimeter of each rectangle to the length of the short side.

    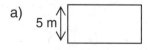 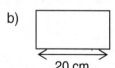  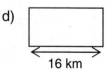

4. Patti says, "100 cm = 1 m, so 100 cm² = 1 m²." Explain why this is wrong.

ME6-30: Area of Parallelograms

1. The rectangle was made by moving the shaded triangle from one end of the parallelogram to the other:

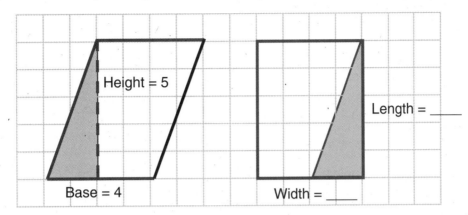

Height = 5

Base = 4

Length = ____

Width = ____

a) Is the area of the rectangle the same as the area of the parallelogram? _____

   How do you know?_____

   _____

b) Fill in the width of the rectangle.

   What do you notice about the base of the parallelogram and the width of the rectangle?

   _____

c) Fill in the length of the rectangle.

   What do you notice about the height of the parallelogram and the length of the rectangle?

   _____

d) Recall that, for a rectangle: Area = length × width.

   Can you write a formula for the area of a parallelogram using the base and height?

2. Measure the height of the parallelograms using a protractor and a ruler.
   Measure the base using a ruler.
   Find the area of the parallelogram using your formula from Question 1 d) above.

   a) 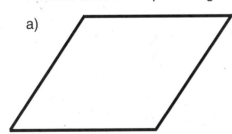   b)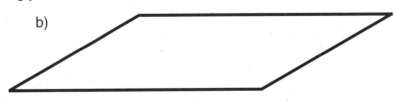

3. Find the area of the following parallelograms.

   a) Base = 5 cm   b) Base = 4 cm   c) Base = 8 cm   d) Base = 3.7 cm
      Height = 7 cm   Height = 3 cm   Height = 6 cm   Height = 6 cm

ME6-31: Area of Triangles

1. a) Draw a dotted line to show the height of the triangle.

   Then find the length of the height and the base of the triangles (in cm).

   The first has been done for you.

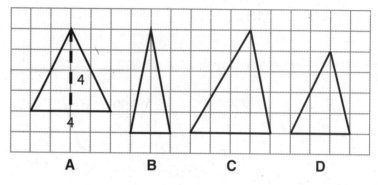

A   B   C   D

   b) Find the area of each triangle above by dividing it into two right angle triangles.

   Area of A: _____   Area of B: _____

   Area of C: _____   Area of D: _____

   REMEMBER:

   Area of Triangle = Area of Rectangle divided by 2

2. Parallelogram B was made by joining two copies of Triangle A together. How can you find the area of Triangle A?

   HINT: Use what you know about the area of parallelograms.

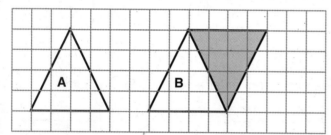

   _____

   _____

   _____

3. Find the area of the triangle by joining two copies of the triangle together to form a parallelogram as in Question 2.

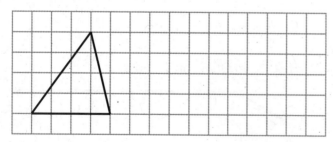

4. Write a formula for the area of a triangle using the base and the height of the triangle.

   HINT: How are the areas of the triangles in Questions 2 and 3 related to the areas of the parallelograms?

   _____

5. Show how you would calculate the area of Triangle A in Question 1 using your formula.

   _____

ME6-32: Investigations

1. On the previous page, you discovered the formula:

   Area of Triangle = (base × height) ÷ 2

   Find the area of a triangle with the dimensions.

   a) Base = 6 cm

      Height = 2 cm

      Area =

   b) Base = 4 cm

      Height = 3 cm

      Area =

   c) Base = 6 cm

      Height = 4 cm

      Area =

   d) Base = 3.2 cm

      Height = 8 cm

      Area =

2. Previously, you discovered the formula:

   Area of a Parallelogram = base × height

   Find the area of a parallelogram with the dimensions.

   a) Base = 5 cm

      Height = 7 cm

      Area =

   b) Base = 10 cm

      Height = 17 cm

      Area =

   c) Base = 3.5 cm

      Height = 9 cm

      Area =

   d) Base = 2.75 cm

      Height = 8 cm

      Area =

3. Measure the base and height of the triangle using a ruler. Then find the area of the triangle.

   a)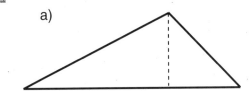

   b)

   c)

4. Find the area of each shape by subdividing it into triangles and rectangles.

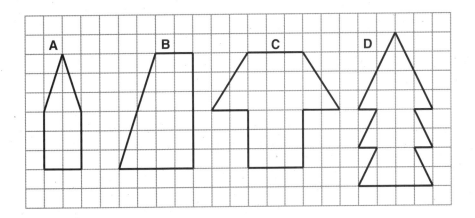

5.

   Draw a line to cut the figure into two rectangles.

   Calculate the area of the two rectangles and add the areas to get the area of the figure.

jump math
MULTIPLYING POTENTIAL.

6. Find the measurements of the sides that are not labelled.
    Then calculate the perimeter and area of each figure.
    CAREFUL: Not all sides have been provided with measurements.

a)

2

6

4

6

Perimeter: _____

Area: _____

b)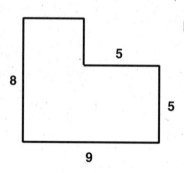

8

5

5

9

Perimeter: _____

Area: _____

7.

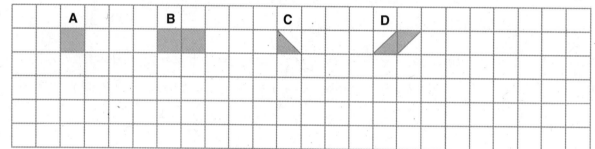

a) Two polygons are <u>similar</u> if they are the same shape. Draw a shape similar to the original, making each base two times as long. How high should you make the new shape?

b) Find the area (in square units) of each original shape. Then find the area of each new shape.

Area of A: _____   Area of B: _____   Area of C: _____   Area of D: _____

Area of the new shape: Area of the new shape: Area of the new shape: Area of the new shape:

_____   _____   _____   _____

c) When the base and the height of a shape are doubled, what happens to the area of the shape?

_____

8. A square has area 25 cm^2. What length is each side? What is its perimeter?

9. A rectangle has area 12 cm^2 and length 6 cm. What is its width? What is its perimeter?

10. A parallelogram has base 10 cm and area 60 cm^2. How high is the parallelogram?

11. Draw a rectangle on grid paper. Draw a second rectangle with sides that are twice as long.
     Is the perimeter of the larger rectangle 2 times or 4 times the perimeter of the smaller rectangle?

12. On grid paper, draw two different rectangles.
     Make the one with the smaller area have the greater perimeter.

Answer the following questions in your notebook.

13. Each square on the grid represents an area of 25 cm².

    What is the area of each figure?

    How do you know?

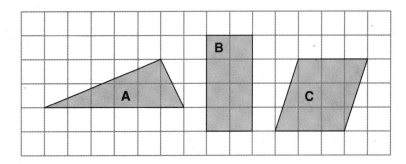

14. Each edge on the grid represents 0.5 cm.

    Is the perimeter of the rectangle greater than or less than 14.5 cm?

    How do you know?

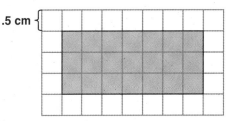

15. The picture shows plans for two parks.

    What is the perimeter of each park?

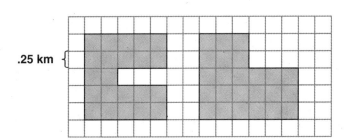

16.  What fraction of the area of the rectangle is the triangle?

    How do you know?

17. 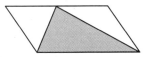 What fraction of the area of the parallelogram is the area of the triangle?

    How do you know?

18. The area of the shaded triangle is 8 m². What is the perimeter of the square?

    How do you know?

19. The area of a triangle is 20 cm², and its base is 10 cm.

    What is the height of the triangle? How can you check your answer?

20. Alex is doing a science project on swimming pools. What could he measure using …

    a) metres (m)?   b) metres squared (m²)?   c) metres cubed (m³)?

    d) kilograms (kg)?   e) litres (L)?   f) kilometres per hour (km/h)?

The capacity of a container is how much it can hold. Capacity is measured in mL, L, etc.
For instance, the capacity of a large bottle of water is 1 L.

Volume, which is measured in cm^3, m^3, etc., is related to capacity:

1 mL is equivalent to 1 cm^3 and   1 L = 1000 cm^3

----

1. Audrey places a layer of centicubes in the bottom of a small glass box.

    a) How many centicubes are in the box now? _____

    b) What is the volume occupied by one layer of centicubes?

        _____

    c) Write a multiplication statement for the volume
        of one layer of centicubes:

        _____

    d) Write a multiplication statement for the
        volume that would be occupied if Audrey
        placed <u>two</u> layers of centicubes in the box:

        _____

    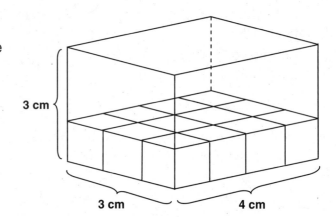

    3 cm

    3 cm   4 cm

    e) Write a multiplication statement for the volume of <u>three</u> layers of centicubes: _____

    f) What is the volume of the glass box? _____

    g) What is the capacity of the box? _____

2. a) Write a multiplication statement for the number of centicubes in each picture.

    (i) 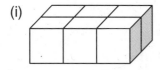   (ii) 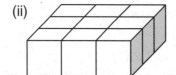   (iii)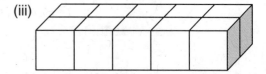

    b) Each picture in part a) above shows the number of centicubes needed to cover the base
        of a box that is 5 cm high.
        Write a multiplication statement for the volume of each box.

    c) If you know the length, width and height of a box with a rectangular base, how do you
        calculate its capacity?

3. Write one possible set of lengths, widths and heights for a box with the capacity …

    a) 12 mL   b) 8 mL   c) 18 mL   d) 24 mL

ME6-34: Kilograms and Metric Tonnes

1 tonne = 1 000 kg

Since the average mass of a soccer player is 100 kg, it would take about 10 soccer players to make a tonne:

100 kg + 100 kg + 100 kg + 100 kg + 100 kg + 100 kg + 100 kg + 100 kg + 100 kg + 100 kg = 1000 kg = 1 tonne

----

1. Multiply the numbers below by 1 000 by moving the decimal 3 places to the right.

   a) $1\,000 \times 2.700$ = ___2 700___   b) $48 \times 1\,000$ = ___4 800___   c) $6 \times 1\,000$ = _____

   d) $1\,000 \times 3.4$ = _____   e) $1\,000 \times 8.1$ = _____   f) $1\,000 \times 1.2$ = _____

   g) $1\,000 \times 9.8$ = _____   h) $1\,000 \times 4.05$ = _____   i) $1\,000 \times 2.3$ = _____

2. Convert the following measurements in tonnes into kg.
   HINT: Multiply by 1000.

   a) 5 t =   b) 18 t =   c) 6 t =   d) 50 t =

   e) 1.5 t =   f) .31 t =   g) 45.5 t =   h) 26 t =

3. An average 11-year-old child weighs about 40 kg.

   a) Skip count by 40 to 1 000: __40__ , __80__ , ____ , ____ , ____ , ____ , ____ , ____ , ____ , ____ ,

      ____ , ____ , ____ , ____ , ____ , ____ , ____ , ____ , ____ , ____ , ____ , ____ , ____ , ____

   b) How many 40s make 1 000? _____

   c) About how many 11-year-olds would weigh 1 tonne? _____

4. Twenty-five friends plan to take a river rafting trip. The average mass of the friends is 50 kg. The raft can carry 1 tonne. Can all the people ride on the raft?

5. An African elephant has a mass of 5000 kg. How many tonnes is this?
   Elephants eat 150 kg of food per day.
   Would an elephant eat more or less than 1 tonne in a week?

6. A moving van can hold 1.2 tonnes. Can Merlinda's family move all of their furniture in one trip?

 150 kg   40 kg   150 kg   75 kg   100 kg   50 kg   70 kg   250 kg   50 kg   65 kg

The different ways an event can happen are called **outcomes** of the event.

When Alice plays a game of cards with a friend, there are three possible outcomes: Alice (1) wins, (2) loses or (3) the game ends without a winner or a loser (this is sometimes called a tie or a draw).

REMEMBER: **A coin has two sides, heads and tails.**   **A die has six sides, numbered 1 to 6.**

----

1. What are the possible outcomes when …

    a) you flip a coin? _____$\frac{1}{2}$_____

    b) you roll a die (a cube with numbers from one to six on its faces)? ___$\frac{1}{6}$___

    c) a hockey team plays a championship match? ___$\frac{1}{12}$___

2. How many different outcomes are there when you:

    a) roll a die? __6__   b) flip a coin? __2__   c) play chess with a friend? __16__

3. What are the possible outcomes for these spinners? The first one is done for you.

    a)   b)   c)   d)

    ____You spin a 1, 2,____   ___you spin___   ___you spin___   ___you spin___
    _____3 or 4_____   ____6____   ___2,3,4___   ___6,9___
    __4__ outcomes   __1__ outcomes   __3__ outcomes   __2__ outcomes

4. lie

    You draw a marble from a box. How many different outcomes are there in each of the following cases?

    a)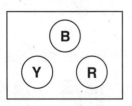

    __1/3__ outcomes

    b)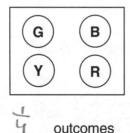

    __1/4__ outcomes

5. List all the outcomes that are …

    a) even numbers:
        $\frac{3}{8}$

    b) odd numbers:
        $\frac{4}{8}$

    c) greater than 9:
        $\frac{1}{8}$

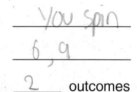

PDM6-22: Probability

Fractions can be used to describe **probability**.

$\frac{3}{4}$ of the spinner is red, so the probability of spinning red is $\frac{3}{4}$.

There are 3 ways of spinning red and 4 ways of spinning any colour (either red or green).

The fraction $\frac{3}{4}$ compares the number of chances of spinning red (3, the numerator) to the number of chances of spinning any colour (4, the denominator).

---

1. For each of the following situations, how many ways are there of …

a)

drawing a green marble?

1/3

drawing a marble of any colour?

$\frac{1}{3}$

b)

drawing a red marble?

1/4

drawing a marble of any colour?

$\frac{3}{4}$

c)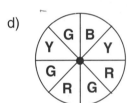

spinning green?

$\frac{2}{4}$

spinning any colour?

$\frac{3}{4}$

d)

spinning green?

$\frac{3}{4}$

spinning any colour?

$\frac{6}{8}$

2. For each spinner, what's the probability (P) of spinning red?

P(Red) =	# of ways of spinning red
	# of ways of spinning any colour

a)

P(Red) = $\frac{2}{5}$

b)

P(Red) = $\frac{1}{2}$

c)

P(Red) = $\frac{1}{4}$

d)

P(Red) = $\frac{1}{3}$

3. What is the probability of throwing a dart and hitting blue? Reduce your answer if possible.

a)

B	R
G	B

P(Blue) = $\frac{2}{4}$

b)

B	R	G

P(Blue) = $\frac{1}{3}$

c)

B	R	R
G	B	Y

P(Blue) = $\frac{2}{6}$

d)

R	B	G
Y		B

P(Blue) = $\frac{2}{5}$

4. For each spinner, write the probability of the given events.
   HINT: Cut the spinners into equal parts.

a)

P(Blue) = $\frac{2}{6}$

b)

P(Red) = $\frac{1}{4}$

c)

P(Yellow) = $\frac{3}{4}$

d)

P(Green) = $\frac{3}{8}$

REMEMBER: A die has the numbers from 1 to 6 on its faces.

5. a) List the numbers on a die:

b) How many outcomes are there when you roll a die?

_____

_____

6. a) List the numbers on a die that are even:

b) How many ways can you roll an even number on a die?

c) What is the probability of rolling an even number on a die?

_____

_____

_____

7. a) List the numbers on a die that are greater than 4:

b) How many ways can you roll a number greater than 4?

c) What is the probability of rolling a number greater than 4 on a die?

_____

_____

_____

8. a) List the numbers on a die that are less than 3:

b) List the numbers on a die that are odd:

c) List the numbers on a die that are multiples of 3:

_____

_____

_____

What is the probability of rolling a number less than 3 on a die?

What is the probability of rolling an odd number on a die?

What is the probability of rolling a multiple of 3 on a die?

_____

_____

_____

9. 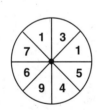   Write a fraction that gives the probability of spinning …

a) the number 1.   b) the number 3.

c) an even number.   d) an odd number.

e) a number less than 5. f) a number greater than 5.

10.    Write a fraction that gives the probability of spinning …

a) the letter A.   b) the letter C.

c) the letter E.   d) a vowel.

e) a consonant.   f) a letter that appears in the word "Canada."

11. Clare says the probability of rolling a 5 on a die is $\frac{5}{6}$. Emma says the probability is $\frac{1}{6}$. Who is right? Explain.

12. Design a spinner on which the probability of spinning red is $\frac{3}{8}$.

PDM6-23: Further Probability

NOTE: When two or more events have the same chance of occurring, the events are <u>equally likely</u>.

1. a) Are your chances of spinning red and yellow equally likely? Explain.

   b) Are your chances of spinning red and yellow equally likely? Explain.

2. A game of chance is **fair** if both players have the same chance of winning.
   Which of the following games are fair?
   For the games that aren't fair, which player has the better chance of winning?

   a)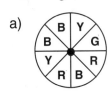

   Player 1 must spin red to win.

   Player 2 must spin blue to win.

   Is it fair?   Y   N

   b)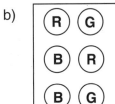

   Player 1 must draw red to win.

   Player 2 must draw blue to win.

   Is it fair?   Y   N

3. Imogen throws a dart at the board. Write the probability of the dart landing on each colour.

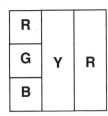

   P(R) = _____

   P(G) = _____

   P(Y) = _____

   P(B) = _____

4. Write letters A, B, and C on the spinner so that the probability of spinning …

   ➢ an A is .3
   ➢ a B is .5
   ➢ a C is .2

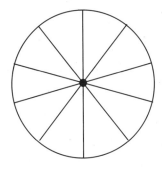

5. a) How many children are in the class?

   b) A child is picked to make the morning announcement. What is the probability the child is a girl?

   c) What is the probability the child is a 10-year-old boy?

   d) Make up your own problem using the numbers in the chart.

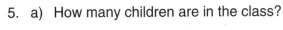

Age	Number of Boys	Number of Girls
10	3	2
11	4	7
12	5	4

PDM6-24: Expectation

Kate plans to spin the spinner 15 times to see how many times it will land on yellow.

Since $\frac{1}{3}$ of the spinner is yellow, Kate **expects** to spin yellow $\frac{1}{3}$ of the time.

Kate finds $\frac{1}{3}$ of 15 by dividing by 3:   **15 ÷ 3 = 5**

So she expects the spinner to land on yellow <u>five</u> times.

NOTE: The spinner may not actually land on yellow five times, but five is the most likely number of "yellow" spins.

-----

1. Write the number of pieces in the pie and the number of pieces shaded.
   Then circle the pies where <u>half</u> the pieces are shaded.

   a) 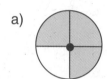   b)    c) 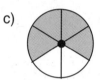   d) 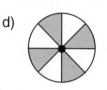   e)

   ___ pieces shaded   ___ pieces shaded   ___ pieces shaded   ___ pieces shaded   ___ pieces shaded

   ___ pieces   ___ pieces   ___ pieces   ___ pieces   ___ pieces

2. Circle the pies where half the pieces are shaded.
   Put a large 'X' through the pies where less than half the pieces are shaded.
   HINT: Count the shaded and unshaded pieces first.

   a)   b)   c)   d)   e)   f)

 3. Using long division, find …

   a) $\frac{1}{2}$ of 10   b) $\frac{1}{2}$ of 24   c) $\frac{1}{2}$ of 48   d) $\frac{1}{2}$ of 52

4. 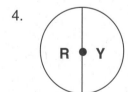   What fraction of your spins would you expect to be red?

   a) I would expect _____ of the spins to be red.

   b) If you spun the spinner 20 times, how many times would you expect to spin red? _____

5. If you flip a coin 40 times, how many times would you expect to flip heads? Explain.

6. Use long division to find the following.

   a) $\frac{1}{3}$ of 42   b) $\frac{1}{3}$ of 75   c) $\frac{1}{4}$ of 52   d) $\frac{1}{4}$ of 84

MULTIPLYING POTENTIAL

Probability & Data Management 2

PDM6-24: Expectation (continued)

7. How many times would you expect to spin yellow if you spun the spinner …

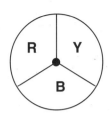

   a) 15 times?　　　b) 36 times?　　　c) 66 times?

   _____　　　_____　　　_____

8.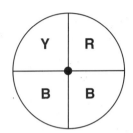

   How many times would you expect to spin red if you spun the spinner …

   a) 16 times?　　　b) 44 times?　　　c) 96 times?

   _____　　　_____　　　_____

9. Colour the marbles red and green (or label them using R and G) to match the probability of drawing a marble of the given colour.

   a) P(Red) = $\frac{1}{2}$　　　b) P(Green) = $\frac{1}{3}$　　　c) P(Red) = $\frac{2}{5}$

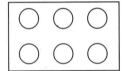 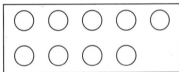 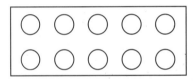

   d) P(Red) = $\frac{3}{4}$　　　e) P(Green) = $\frac{2}{3}$　　　f) P(Red) = $\frac{3}{4}$

 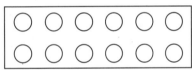 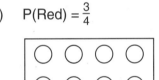

10. Sketch a spinner on which you would expect to spin red $\frac{3}{4}$ of the time.

11. On a spinner, the probability of spinning yellow is $\frac{2}{3}$.

    What is the probability of spinning a colour that is not yellow? Explain your answer with a picture.

12. How many times would you expect to spin blue if you used the spinner 50 times?

    Explain your thinking.

13. How many times would you expect to spin yellow if you used the spinner 100 times?

    Explain your thinking.

Probability & Data Management 2

PDM6-25: Describing Probability

- If an event cannot happen it is **impossible**.
  Example: Rolling the number 8 on a die is impossible (since a die only has the numbers 1, 2, 3, 4, 5, and 6 on its faces).

- If an event must happen it is **certain**.
  Example: When you roll a die it is certain that you will roll a number less than 7.

- It is **likely** that you would spin yellow on the spinner shown (since more than half the area of the spinner is yellow).

- It is **unlikely** that you would spin red on the spinner shown (since there is only a small section of the spinner that is red).

- When an event is expected to occur exactly half the time, we say that there is an **even** chance of the event occurring.

----

1. Describe each event as "likely" or "unlikely."
   HINT: Start by finding out if the event will happen more than half the time or less than half the time.

   a)

   spinning red is:

   _____

   b)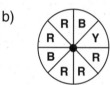

   spinning red is:

   _____

   c)

   spinning green is:

   _____

   d)

   spinning green is:

   _____

2. Beside each even, write **L** for Likely, **U** for Unlikely, or **E** for Even.

   a) 8 socks in a drawer; 5 black socks
      Event: You pull out a black sock. _____

   b) 20 coins in a pocket; 9 pennies
      Event: You pull out a penny. _____

3. Describe each event as "impossible", "unlikely", "likely" or "certain."

   a)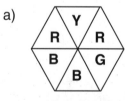

   spinning blue:

   _____

   b)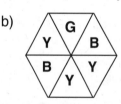

   spinning red:

   _____

   c)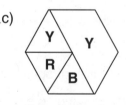

   spinning yellow:

   _____

   d)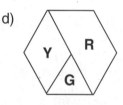

   spinning yellow:

   _____

4. Match the probability with the description of the event.

   A. The probability of the event is 0.

   B. The probability of the event is 1.

   C. The probability of the event is less than $\frac{1}{2}$.

   D. The probability of the event greater than $\frac{1}{2}$.

   _____ the event is unlikely

   _____ the event is impossible

   _____ the event is likely

   _____ the event is certain

Probability & Data Management 2

PDM6-25: Describing Probability (continued)

NOTE: You can show the likelihood of events using a probability line.

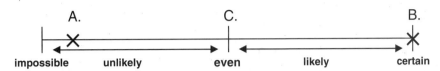

A. It could snow in Toronto in August but it is very unlikely.
So you would mark that event near impossible on the probability line.

B. If you roll a die, you will certainly get a number less than 19.
So you would mark that event certain on the probability line.

C. **Even** probability means the event will occur half the time.

5.

| impossible | unlikely | even | likely | certain |

Mark a point on the line above to show ...

A. The chance of rolling a number less than 20 on a die.

B. The chance of seeing a wolf on the street.

C. The chance of flipping tails on a coin.

D. The chance of rolling a number greater than 2 on a die.

6. Mark an 'X' on the number line to show the probability of spinning red (R), green (G), yellow (Y), and blue (B). Be sure to label the 'X' with the letter of the colour.

0   1

7.

Which colour are you most likely to spin? _____

Which two colours are you least likely to spin? _____

Which word best describes your chances of spinning red?

      Unlikely   Even   Likely

Which word best describes your chances of spinning green?

      Unlikely   Even   Likely

8.

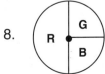

Is each outcome on the spinner equally likely? Explain.

Probability & Data Management 2

1.
Tanya and Daniel play a game of chance with the spinner shown.

➤ If it lands on yellow, Tanya wins.

➤ If it lands on red, Daniel wins.

a) Tanya and Daniel play the game 20 times.

How many times would you <u>predict</u> that the spinner would land on red? _____

b) When Tanya and Daniel play the game, they get the results shown in the chart.

Daniel says the game isn't fair. Is he right? Explain.

Green	Red	Yellow
卌 丨	丨丨丨丨	卌 卌

_____

_____

_____

_____

2. If you flip a coin 20 times ...

a) How many of your flips would you expect to be heads?

b) Which chart shows the result you would most likely get?

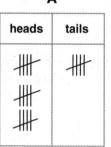
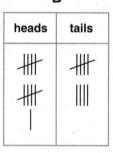

A		**B**		**C**	
heads	tails	heads	tails	heads	tails
卌 卌 卌	卌	卌 卌 丨	卌 丨丨丨丨	卌 丨丨	卌 卌 丨丨丨

3. If you spun the spinner 18 times ...

a) How many of your spins would you expect to be green?

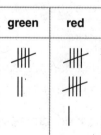

b) Which of the charts shows a result you would be most likely to get?

c) Which result would surprise you?

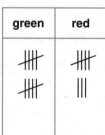
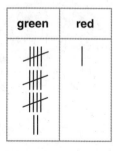

A		**B**		**C**	
green	red	green	red	green	red
卌 丨丨	卌 卌 丨	卌 卌	卌 丨丨丨	卌 卌 卌 丨丨	丨

4. Place the point of your pencil inside a paper clip in the middle of the spinner.
   NOTE: Be sure to hold the pencil still so you can spin the paper clip around the pencil.

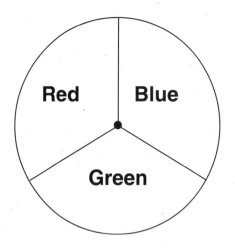

   a) If you spin the spinner 30 times, how many times would you predict spinning red? Show your work.
      HINT: Think of dividing 30 spins into three equal parts.

   b) Spin the spinner 30 times. Make a tally of your results. Did your results match your expectations?

5. If you roll a die repeatedly, what fraction of the time would you expect to roll a 6? Explain.

6. You have 3 bills in your pocket:
   - a $5 bill
   - a $10 bill
   - a $20 bill

   You reach in and pull out a <u>pair</u> of bills.

   a) What are all the possible combinations of two bills you could pull out?

   b) Would you expect to pull a pair of bills that add up to $30? Are the chances likely or unlikely?

   c) How did you solve the problem in part b)? Did you use a list? A picture? A calculation?

7. Write numbers on the spinners to match the probabilities given.

   a)

   The probability of spinning a 3 is $\frac{1}{4}$.

   b)

   The probability of spinning an even number is $\frac{5}{6}$.

   c)

   The probability of spinning a multiple of 3 is $\frac{2}{5}$.

   d)

   The probability of spinning a 2 is $\frac{1}{2}$.

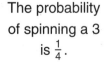 jump math
MULTIPLYING POTENTIAL.

Probability & Data Management 2

1. Match the spinner to the correct statement.

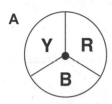

A   B   C   D

_____ Spinning blue is three times as likely   _____ Spinning any colour is equally likely.
as spinning white.

_____ Spinning blue and white are equally likely.   _____ Spinning one colour is twice as
likely as spinning any other colour.

2. Match the net for a tetrahedron ( ) to the correct statement.

A   B   C   D

_____ The probability of rolling a 1 is $\frac{1}{4}$.   _____ The probability of rolling an odd number is $\frac{3}{4}$.

_____ The probability of rolling an even   _____ The probability of rolling a 3 is $\frac{1}{2}$.
number is $\frac{3}{4}$.

3. Match the net for a cube ( ) to the correct statement.

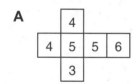 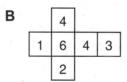 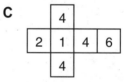 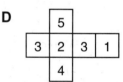

A   B   C   D

_____ The probability of rolling a 5 is $\frac{1}{6}$.   _____ The probability of rolling a 4 is $\frac{1}{2}$.

_____ The probability of rolling an even   _____ The probability of rolling a 1 is the same
number is $\frac{1}{2}$.   as the probability of rolling a 3.

4. Write numbers on the spinners to match the probabilities. The probability of spinning ...

a)   b)   c)   d)

  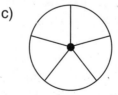 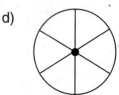

... a 3 is $\frac{1}{2}$.   ... an even number   ... a multiple of 3 is $\frac{2}{5}$.   ... a 2 is $\frac{1}{2}$.
is $\frac{5}{6}$.

PDM6-28: Tree Diagrams

TEACHER NOTE: The next three sections are enriched units, beyond the regular curriculum.

At sports camp, Erin has these choices of sports:

Morning – gymnastics or rowing

Afternoon – volleyball, hockey or rugby

Erin draws a **tree diagram** so she can see all of her choices.

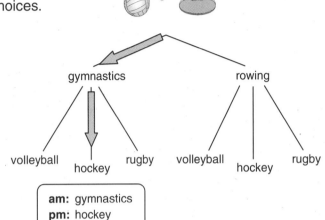

Step 1: She writes the name of her two morning choices at the end of two branches.

Step 2: Under each of her morning choices, she adds three branches (one for each of her three afternoon choices).

Step 3: Following any path along the branches (from the top of the tree to the bottom), you will find one of Erin's choices:

> **am:** gymnastics
> **pm:** hockey

Example: The path highlighted by arrows shows gymnastics in the morning and hockey in the afternoon.

---

1. Follow a path from the top of the tree to a box at the bottom and write the sports named on the path in the box. Continue until you have filled in all the boxes.

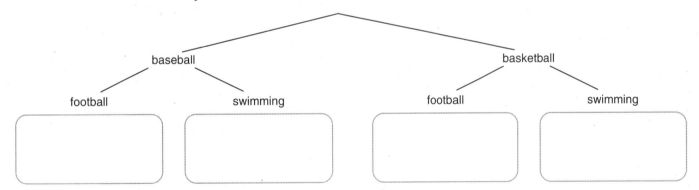

2. Complete the tree diagram to show all of the possible outcomes from flipping a coin twice (H = heads and T = tails).

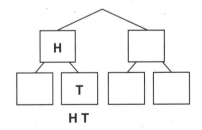

3. Matthew's camp offers the following choices of activities:

   Morning – cricket or rowing
   Afternoon – tae kwon do or judo.

   Draw a tree diagram (like the one in Question 1) to show all of his choices.

4. Emma is playing a role-playing game and her character is exploring a tunnel in a cave.

   List all the paths through the cave, using U for up and D for down.

   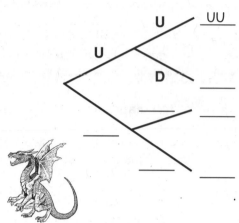

   a) How many paths are there through the cave?

   _____

   b) At the end of one path, a dragon is waiting.

   Do you think it's likely or unlikely that Emma's character will meet the dragon? Explain.

   _____

   _____

5. Complete the tree diagram to show all of the possible outcomes from first flipping a coin and then drawing a marble from the box.

    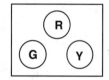

   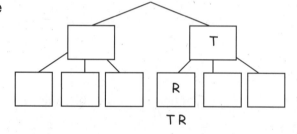

6.  Draw a tree diagram to show all the combinations of numbers you could spin on the two spinners.

   a) How many pairs of numbers add to four?

   b) How many pairs of numbers have a product of four?

7. A restaurant offers the following choices for breakfast:

   Main Course – Eggs or Pancakes

   Juice – Apple, Orange or Grape

   Draw a tree diagram to show all the different breakfasts you could order.

8. Make a tree diagram to show all the combinations of points you could get throwing two darts.

   How many combinations add to 5?

   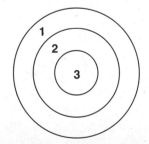

PDM6-29: Counting Combinations

Abdul wants to know how many outcomes there are for a game with <u>two</u> spinners.

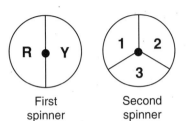

First spinner

Second spinner

	First spinner	Second spinner
Step 1: There are <u>3 outcomes</u> on the second spinner, so Abdul lists each colour on the first spinner <u>3 times</u>.	R	
	R	
	R	
	Y	
	Y	
	Y	

	First spinner	Second spinner
Step 2: Beside each colour, Abdul writes the <u>3 possible outcomes</u> on the second spinner.	R	1
	R	2
	R	3
	Y	1
	Y	2
	Y	3

The list shows that, altogether, there are <u>6 outcomes</u> for the game.

For each question below, answer parts a) and b) first.
Then complete the list of combinations to show all the ways Abdul can spin a colour and a number.

1.

a) How many outcomes are on the second spinner?

_____

b) How many times should Abdul write B (for blue) and R (for red) on his list?

_____

c) How many outcomes does this game have altogether?

_____

First spinner	Second spinner

2.

a) How many outcomes are on the second spinner?

_____

b) How many times should Abdul write Y (for yellow) and G (for green) on his list?

_____

c) How many outcomes does this game have altogether?

_____

First spinner	Second spinner

3.

a) How many outcomes are on the second spinner?

_____

b) How many times should Abdul write G (for green), B (for blue) and Y (for yellow) on his list?

_____

c) How many outcomes does this game have altogether?

_____

First spinner	Second spinner

4.

a) How many outcomes are on the second spinner?

_____

b) How many times should Abdul write G (for green), B (for blue) and Y (for yellow) on his list?

_____

c) How many outcomes does this game have altogether?

_____

First spinner	Second spinner

jump math
MULTIPLYING POTENTIAL.

Probability & Data Management 2

5. If you flip a coin there are two outcomes: heads (H) and tails (T).

   List all the outcomes for flipping a coin and spinning the spinner.

Coin	Spinner

6. Peter has a quarter and a dime in his left pocket, and a dime and a nickel in his right pocket.

   He pulls one coin from each pocket.

   List all the combinations of coins that he could pull out of his pockets.

Right pocket	Left pocket	Value of the coins

7. Clare can choose the following activities at art camp:

   Morning – painting or music

   Afternoon – drama, pottery or dance

   She makes a chart so she can see all of her choices. She starts by writing each of her morning choices 3 times.

   a) Complete the chart to show all of Clare's choices.

   b) Why did Clare write each of her choices for the morning 3 times?

Morning	Afternoon
painting	
painting	
painting	
music	
music	
music	

8. Make a chart to show all the activities you could choose at a camp that offered the following choices:

   Morning – swimming or tennis   **Afternoon** – canoeing, baseball or hiking

9.

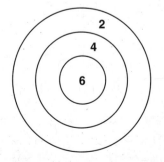

1st dart	2nd dart	Total Score

   a) Record all scores you could get by throwing two darts at the dart board. (Assume both darts land on the board.)

   b) Are there any combinations that give the same score?

PDM6-30: Compound Events

1. Write a set of ordered pairs to show all the combinations you could spin on the two spinners.
   NOTE: The first one has been done for you, and the next two have been partially done.

   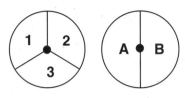

   (_1_ , _A_)   (___ , _A_)   (___ , _A_)

   (___ , ___)   (___ , ___)   (___ , ___)

   a) How many outcomes are there? _____

   b) How many ways can you spin... (i) a 1 on the first spinner and an A on the second? _____

   (ii) an odd number on the first and a B on the second? _____

   c) State the <u>probability</u> of spinning each situation in part b): (i) _____   (ii) _____

2. a) Write a set of ordered pairs to show all the combinations you could spin on these two spinners.

   b) State the probability of spinning ...

   (i) a 1 on the first spinner and an A on the second.

   (ii) an odd number on the first and a B on the second.

3. Write a set of ordered pairs to show all the combinations for the following pairs.

   a) Flipping a coin and spinning a spinner:

   b) Rolling a pair of tetrahedral dice:

4. Jason has a $5 bill and a $10 bill in his <u>right</u> pocket, and a $5 bill and a $10 bill in his <u>left</u> pocket.

   He pulls <u>one</u> bill from each of his pockets.

   a) List the combinations of bills he could pull from his pockets.

   b) What is the probability that he will pull a pair of bills with a value of $15?

Right pocket	Left pocket	Value of bills

G6-21: Coordinate Systems

Rows and columns can be identified by a pair of numbers in a bracket.
The first number gives the column and the second, the row.

(5,3)

column   row

NOTE: Letters may be used instead of numbers, as in Question 2 below.

---

1. Circle the points in the following positions (connecting the dots first, if necessary).

a)

Column 2
Row 1

b)

Column 3
Row 2

c)

Column 3
Row 1

d)

Column 2
Row 2

e)

(2,1)

f)

(3,2)

g)

(1,2)

h)

(2,3)

2. Circle the points in the following positions.

a)

A B C

(A,3)

b)

X Y Z

(Y,B)

c)

(0,2)

d)

(0,0)

3. Put points at the given positions, labelling each point with the letter written beside the ordered pair. The first one has been done for you.

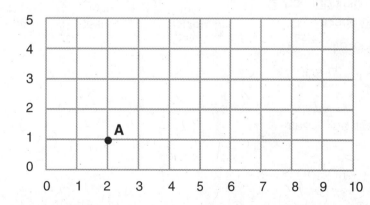

~~A (2,1)~~	B (8,4)	C (5,5)
D (4,2)	E (1,5)	F (10,3)
G (0,0)	H (3,4)	I (6,5)

4. Circle the points in the following positions.

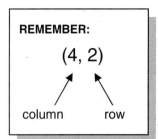

a)

```
4  •  •  •  •
3  •  •  •  •
2  •  •  •  •
1  •  •  •  •
   1  2  3  4
```

Column 2
Row 3

b)

```
Z  •  •  •  •
Y  •  •  •  •
X  •  •  •  •
W  •  •  •  •
   1  2  3  4
```

(2,X)

c)

```
4  •  •  •  •
3  •  •  •  •
2  •  •  •  •
1  •  •  •  •
   1  2  3  4
```

(4,1)

d)

```
4  •  •  •  •  •  •  •
3  •  •  •  •  •  •  •
2  •  •  •  •  •  •  •
1  •  •  •  •  •  •  •
   1  2  3  4  5  6  7
```

(3,4)

e)

```
4  •  •  •  •  •  •  •
3  •  •  •  •  •  •  •
2  •  •  •  •  •  •  •
1  •  •  •  •  •  •  •
   1  2  3  4  5  6  7
```

Column 7
Row 2

f)

```
D  •  •  •  •
C  •  •  •  •
B  •  •  •  •
A  •  •  •  •
   A  B  C  D
```

(A,C)

5. Put points at the given positions, labelling each point with the letter written beside the ordered pair.

A (4,7)   **B** (9,3)

C (2,1)   **D** (2,3)

E (0,5)   **F** (7,7)

G (5,6)   **H** (9,0)

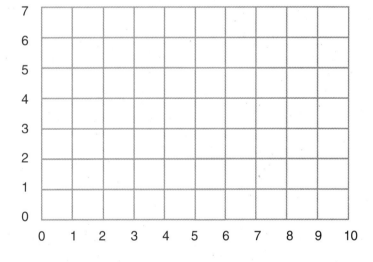

6.

```
C •————————• D
  |         |
B •————————• A
  |         |
  •————————•
(0,0)     (5,0)
```

The diagram to the left shows a grid with some lines left out.

A is at (5,5).

Write the coordinates of B, C and D:

B (,   )   **C** (,   )   **D** (,   )

G6-21: Coordinate Systems (continued)

7. Write the coordinates of the following points.

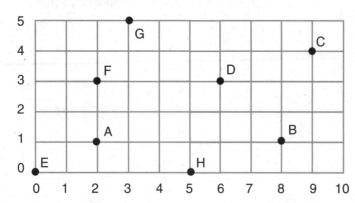

A (,)   B (,)

C (,)   D (,)

E (,)   F (,)

G (,)   H (,)

 TEACHER: Review definitions of polygons before you assign the questions below.

8. Graph each set of ordered pairs and join the dots to form a polygon. Identify the polygon drawn.

a)
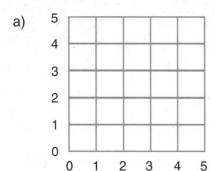

A (1,1)   B (4,1)   C (4,3)

This polygon is a _____.

b)
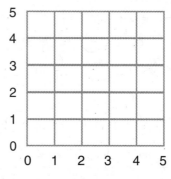

A (1,1)   B (3,1)   C (3,3)   D (2,4)   E(1,3)

This polygon is a _____.

9. Add a dot D so that the four dots form the vertices of ...

a)
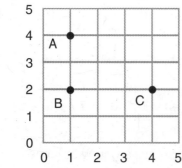

... a rectangle. Then write the coordinates of the four vertices:

A (,) B (,) C (,) D (,)

b)
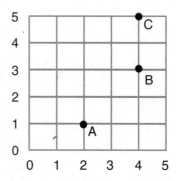

... a parallelogram. Then write the coordinates of the four vertices:

A (,) B (,) C (,) D (,)

BONUS

10. Draw a polygon on grid paper. Tell a friend the coordinates of the vertices of your polygon and see if they can name the polygon.

G6-22: Coordinate Systems (Advanced)

A grid that has been extended to include negative integers is called a **Cartesian coordinate system**.

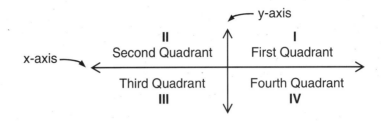

1. Identify and label the origin (0) and the x- and y-axes.

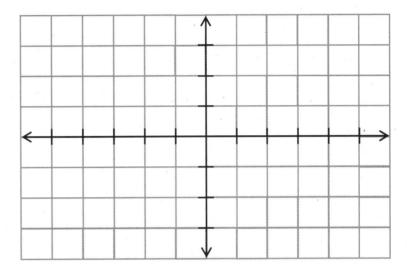

2.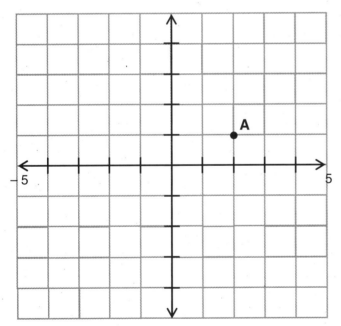

Plot the points.

A (2, 1)   **B** (4, 3)

C (−4, −4)   **D** (3, −5)

E (3, −5)   **F** (5, 0)

G (−3, 0)   **H** (0, −1)

I (0, 4)

3. Plot these points on grid paper and join the dots. What polygon did you make?

a) **A** (2, 2)   **B** (5, 1)   **C** (6, 4)   **D** (3, 5)

b) **A** (−2, 0)   **B** (−2, 4)   **C** (−4, 2)   **D** (−4,−2)

c) **A** (−1,−2)   **B** (0,−2)   **C** (3,−4)   **D** (−1,−4)

d) **A** (2, 1)   **B** (1, 3)   **C** (0, 1)   **D** (1,−1)

 jump math
MULTIPLYING POTENTIAL.

Geometry 2

G6-23: Slides

1. How many units **right** or **left** did the dot slide from position 1 to position 2?

 L

a)

b)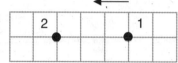

c) 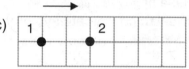 R

_____ units right

_____

_____

2. How many units **right** or **left** and how many units **up** or **down** did the dot slide from position 1 to position 2?

a)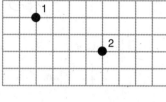

b)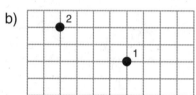

c)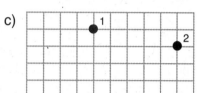

___ units right ___ units down

___ units left ___ units up

___ units right ___ units down

3. Slide the dot.

a) 5 units right; 2 units down

b) 6 units left; 3 units up

c) 3 units left; 4 units down

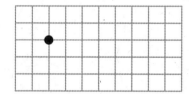

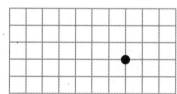

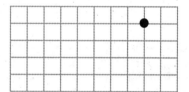

4. Copy the shape into the second grid.

a)

b)

c)

5. Slide the dot three units down, then copy the shape.

a)

b)

c)

d)

6. Slide each figure 5 boxes to the right and 2 boxes down.

a)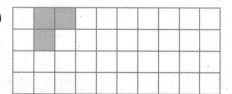

b)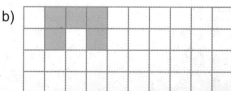

G6-24: Slides (Advanced)

1. The picture shows a translation of a square.

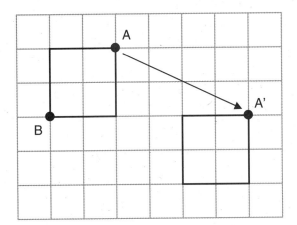

a) Describe how the point A moved to the point A'.

_____

b) Draw an arrow to show where point B moved to under the translation.

c) Describe how point B moved:

_____

d) Did all of the points on the square move by the same amount?

_____

2. Draw a translation arrow from a vertex of shape A to the corresponding vertex in A'.
   Then describe how far the shape slid in moving from position A to A'.

a)
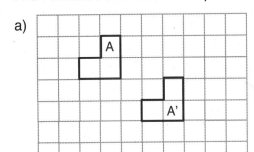

b)
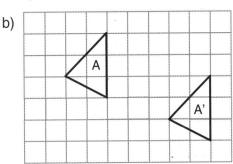

_____   _____

3. Slide the shapes in the grids below. Describe how far the shape moved (right/left and up/down).

a)
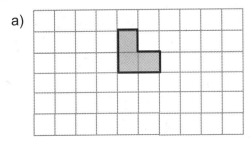

b)
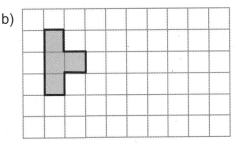

My slide: _____   My slide: _____

4. Translate a shape on grid paper.
   Draw a translation arrow between a point on the shape and a point on the image.
   Describe how far the shape moved (right/left and up/down).

5. Draw 2 coordinate grids on grid paper. Draw the given shapes.
   Translate the shape and write the coordinates of its new vertices.

   a) Square with vertices   A(1,1), B(1,3), C(3,3), D(3,1)   Translate 3 units right, 4 up

   b) Triangle with vertices   A(3,7), B(2,5), C(5,4)   Translate 4 units right, 3 down

Geometry 2

G6-25: Coordinate Systems and Maps

1. Answer the following questions using the coordinate system.

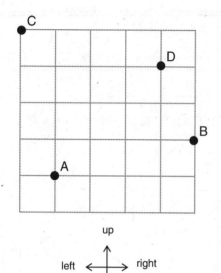

up

left ←→ right

down

a) What point is 1 unit left and 4 up from A?

b) What point is 3 units left and 3 down from D?

c) What point is 3 units down and 5 right of C?

d) Describe how to get from point B to point D:

e) Describe how to go to point B from point A:

f) Describe how to get to point A from point C:

2. Answer the following questions using the coordinate system.

1 km

4				town	
3		hill			cliff
2					valley
1	lake				
	A	B	C	D	E

North

West ←→ East

South

HINT:
Each square represents a square km.

a) What would you find in square (B,3)?

b) What would you find if you travelled 1 km north of the valley?

c) Give the coordinates of the lake.

d) Describe how to get from the town to the hill.

e) Describe how to get from the lake to the cliff.

G6-26: Reflections

Shane reflects the shape by flipping it over the mirror line. Each point on the figure flips to the opposite side of the mirror line, but stays the same distance from the line.

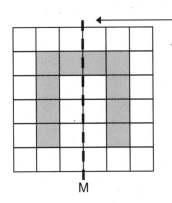

the line of reflection
(or mirror line)

M

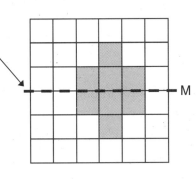

M

---

1. Draw the reflection of the shapes below.

a)
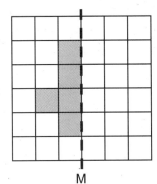
M

b)
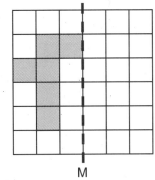
M

c)
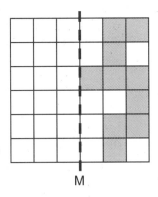
M

2. Draw the reflection, or flip, of the shapes.

a)
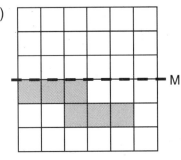
M

b)
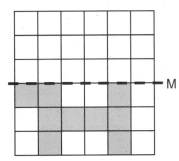
M

c)
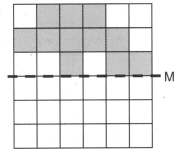
M

3. Draw your own shape in the box below. Now draw the flip of the shape on the other side of the mirror line.

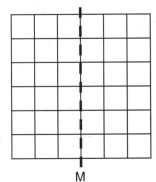
M

BONUS
Are the shapes on either side of the mirror congruent?
Explain your answer.

_____

_____

_____

jump math
MULTIPLYING POTENTIAL.

Geometry 2

G6-26: Reflections (continued)

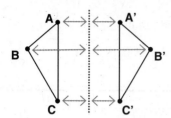

When a point is reflected in a mirror line, the point and the image of the point are the same distance from the mirror line.

A figure and its image are congruent but face in opposite directions.

----

4. Reflect the point P through the mirror line M. Label the image point P'.

a)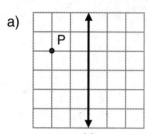

b)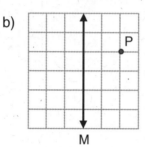

c)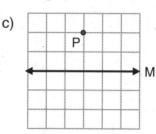

d)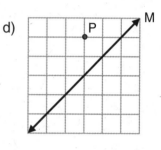

5. Reflect the set of points P, Q and R through the mirror line. Label the image points P', Q' and R'.

a)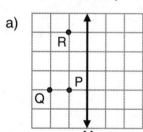

b)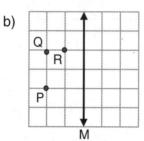

c)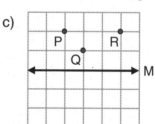

d)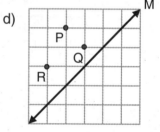

6. Reflect the figure by first reflecting the points on the figure.

a)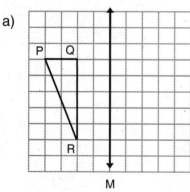

b)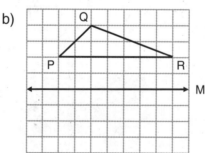

c)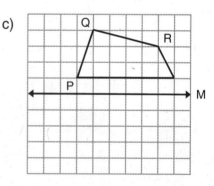

7. Reflect the shapes.

a)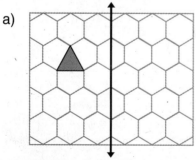

b)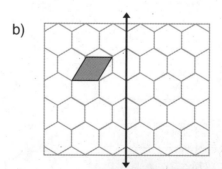

Geometry 2

G6-27: Rotations

Alice wants to rotate this arrow $\frac{1}{4}$ of a turn clockwise:

<u>Step 1:</u>
She draws a circular arrow to show how far the arrow should turn.

<u>Step 2:</u>
She draws the final position of the arrow.

---

1. Write how far each arrow has moved from start to finish.

a)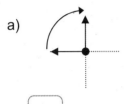

☐ turn clockwise

b)

☐ turn clockwise

c)

☐ turn clockwise

d)

☐ turn clockwise

2. Write how far each arrow has moved counter clockwise from start to finish.

a)

☐ turn counter clockwise

b)

☐ turn counter clockwise

c)

☐ turn counter clockwise

d)

☐ turn counter clockwise

3. Show where the arrow would be after each turn. **HINT: Use Alice's method.**

a)

$\frac{1}{4}$ turn clockwise

b)

$\frac{1}{2}$ turn clockwise

c)

$\frac{3}{4}$ turn clockwise

d)

1 whole turn clockwise

e)

$\frac{1}{2}$ turn
counter clockwise

f)

1 whole turn
counter clockwise

g)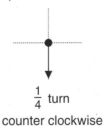

$\frac{1}{4}$ turn
counter clockwise

h)

$\frac{3}{4}$ turn
counter clockwise

BONUS

i)

three $\frac{1}{4}$ turns
counter clockwise

j)

three $\frac{1}{2}$ turns
clockwise

k)

three $\frac{1}{4}$ turns
counter clockwise

l)

two $\frac{3}{4}$ turns
counter clockwise

1. Show what the figure would look like after the rotation. First rotate the dark line, then draw the rest of the figure.

a)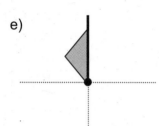

b)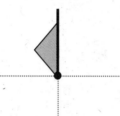

c)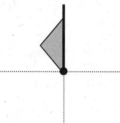

d)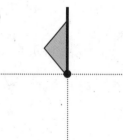

$\frac{1}{4}$ turn clockwise $\frac{1}{2}$ turn clockwise $\frac{3}{4}$ turn clockwise   1 whole turn clockwise

e)

f)

g)

h)

$\frac{1}{4}$ turn clockwise $\frac{3}{4}$ turn clockwise $\frac{1}{4}$ turn counter clockwise $\frac{1}{2}$ turn clockwise

2. Show what the shape would look like after the rotation.

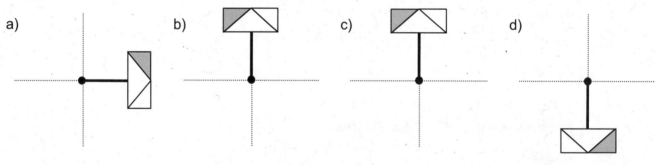

a)

b)

c)

d)

$\frac{1}{4}$ turn clockwise $\frac{3}{4}$ turn clockwise $\frac{1}{4}$ turn counter clockwise $\frac{1}{2}$ turn clockwise

3. Draw a figure on grid paper. Draw a dot on one of its corners. Show what the figure would look like if you rotated it a quarter turn clockwise around the dot.

4. Step 1: Draw a trapezoid on grid paper and highlight one of its sides as shown.

   Step 2: Use a protractor to rotate the line 120° clockwise.

   Step 3: Draw the trapezoid in the new position.

   HINT: You will have to measure the sides and angles of the trapezoid to reconstruct it.

5. Rotate an equilateral triangle 60° clockwise around one of its vertices. What do you notice?

G6-29: Rotations and Reflections

1. Rotate each shape 180° around centre P by showing the final position of the figure.

   Use the line to help you.

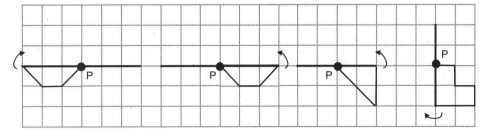

2. Rotate each shape 180° around centre P.
   HINT: First highlight an edge of the figure and rotate the edge (as in Question 1).

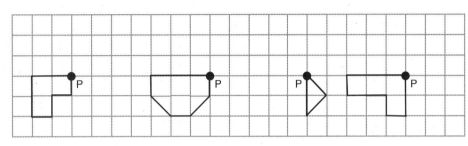

3. Rotate each shape 90° around point P in the direction shown.

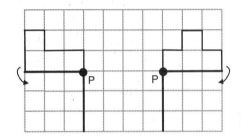

4. Rotate each shape 90° around the point in the direction shown.

   HINT: First highlight a line on the figure and rotate the line 90°.

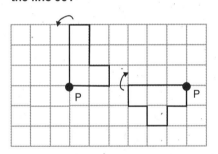

5. Write 90° beside the figure (1 or 2) that was made by rotating the original figure 90° counter clockwise. Then write 180° beside the figure that was made by rotating the original figure 180°.

   a)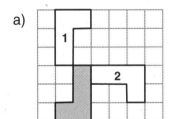

   b)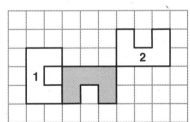

   c)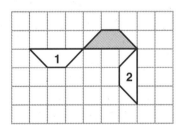

6. Write 180° beside the figure (1 or 2) that was made by rotating the original figure 180°. Then write 'R' on the figure that was made by a reflection. Mark the centre of the rotation and draw a mirror line for the reflection.

   a)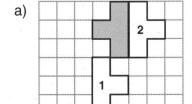

   b)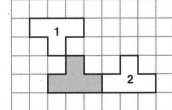

   c)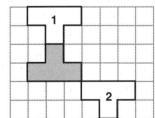

1. Show the image of the figure under each transformation.

a)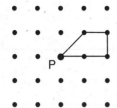

$\frac{1}{4}$ turn clockwise
about point P

b)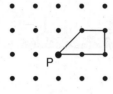

$\frac{1}{2}$ turn clockwise
around point P

c)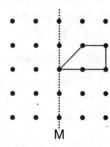

reflection in line M

d)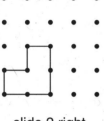

slide 2 right

e)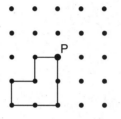

$\frac{1}{4}$ turn counter clockwise
around P

f)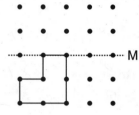

reflection in line M

2. a)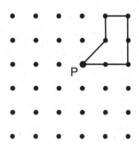

Rotate the figure $\frac{1}{4}$ turn clockwise
around point P.

Then slide the resulting image 2 units left.

b)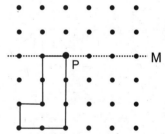

Rotate the figure $\frac{1}{4}$ turn counter clockwise
around point P.

Then reflect the resulting image in the
mirror line.

3. Copy the given figure onto grid paper 3 times.

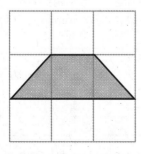

a) Pick any vertex on the figure as a centre of rotation and turn the figure $\frac{1}{4}$ or $\frac{1}{2}$ turn around that vertex. Then slide the figure in any direction. Describe the transformations you used.

b) Rotate the figure around any vertex, then reflect it in a mirror line of your choice.

c) Move the figure by a combination of two transformations. Ask a friend to guess which two transformations you used to move the figure.

G6-31: Transformations (Advanced)

1. a) Translate Figure **A** 4 units left and 2 units down.
       Label the image **B**.

    b) Turn Figure **B** 90° clockwise around point (5,2).
       Label the image **C**.

    c) Write the ordered pairs of the vertices of **C**:

       (_____ , _____) (_____ , _____)

       (_____ , _____)

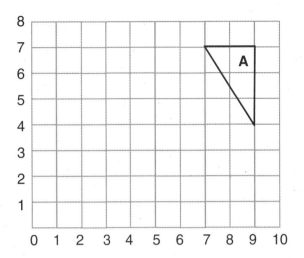

2. a) Reflect the Figure **A** in line **L**.
       Label the image **B**.

    b) Rotate Figure **B** 180° around point (7,2).
       Label the image **C**.

    c) Write the ordered pairs of the vertices of **C**:

       (_____ , _____) (_____ , _____)

       (_____ , _____) (_____ , _____)

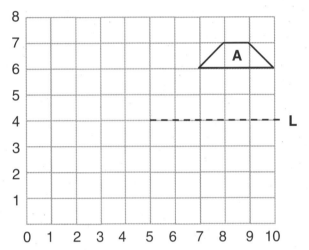

3. Rotate the triangle around the center P by the given angle. Then …

    • Draw a line from point P to the nearest vertex of the triangle.

    • Rotate the line by the given angle, using a protractor (make sure the new line is the same length as the original line).

    • Rotate the triangle as if it was attached to the line (make sure the sides of the new triangle and the original are the same length).

    a) 90° counter clockwise

    b) 90° clockwise

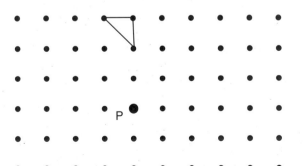

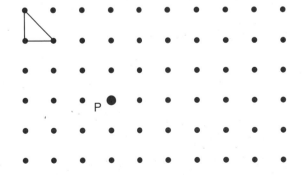

Geometry 2

1.

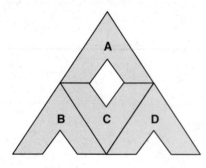

a) Which transformation (slide, reflection or rotation) could you use to move Shape A onto …

   (i) Shape B?   (ii) Shape C?   (iii) Shape D?

b) Philip says: "I can move Shape C onto Shape B using a $\frac{1}{2}$ turn and then a slide." Is he correct?

c) Explain how you could move Shape C onto Shape D using a reflection and a slide.

2. Identify <u>two</u> transformations for which B is the image of A:

3. Describe precisely how each figure moved.

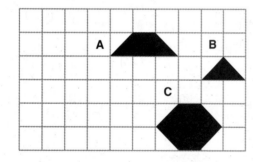  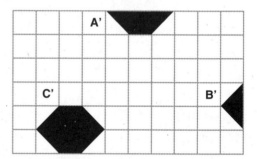

4.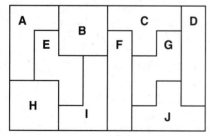

Predict which direction the letter will face after two reflections (through lines L_1 and L_2). Then reflect the letter to test your prediction.

5.

a) Name 5 pairs of congruent shapes in the picture.

b) Draw a mirror line on the picture that would allow you to reflect Shape E onto Shape G.

c) For which shapes can you check congruency using only a slide?

d) For which shapes can you check congruency using …

   (i) a rotation and slide?   (ii) a reflection and a slide?

   Explain.

G6-33: Building Pyramids

TEACHER: For the exercises on this page you will need modeling clay (or plasticine) and toothpicks (or straws).

To make a skeleton for a **pyramid**, start by making a base.
Your base might be a triangle or a square.

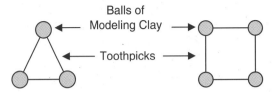

Now add an edge to each vertex on your base and join the edges at a point.

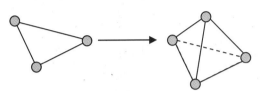

Triangular Pyramid

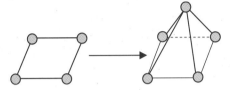

Square Pyramid

After you have made a triangular pyramid and a square pyramid, try to make one with a five-sided base (a pentagonal pyramid). Then fill in the first three rows of the chart below.

- -

1.

	Draw Shape of Base	Number of Sides of Base	Number of Edges of Pyramid	Number of Vertices of Pyramid
Triangular Pyramid				
Square Pyramid				
Pentagonal Pyramid				
Hexagonal Pyramid				

 2. Describe the pattern in each column of your chart.
   Use the pattern to fill in the row for the hexagonal pyramid.

3. Describe any relationships you see between the columns of the chart.
   FOR EXAMPLE: What is the relationship between the number of sides in the base of the pyramid and the number of vertices or the number of edges in the pyramid?

4. How many edges and vertices would an octagonal pyramid have?

Geometry 2

G6-34: Building Prisms

To make a skeleton for a **prism**, start by making a base (as you did for a pyramid). However, your prism will also need a top, so you should make a copy of the base.

Now join each vertex in the base to a vertex in the top.

After you have made a triangular prism and a cube, try to make a prism with two five-sided bases (a pentagonal prism). Then fill in the first three rows of the chart below.

1.

	Draw Shape of Base	Number of Sides of Base	Number of Edges of Prism	Number of Vertices of Prism
Triangular Prism				
Cube				
Pentagonal Prism				
Hexagonal Prism				

2. Describe the pattern in each column of your chart.
    Use the pattern to fill in the row for the hexagonal prism.

3. Describe any relationships you see between the columns of the chart.

4. How many edges and vertices would an octagonal prism have?

G6-35: Edges, Faces, and Vertices

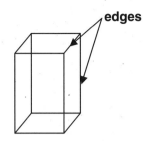

edges

Candice builds a skeleton of
a rectangular prism using wire.

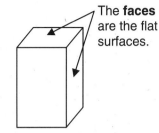

The **faces**
are the flat
surfaces.

She covers the skeleton
with paper.

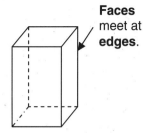

Faces
meet at
edges.

The dotted lines show
the <u>hidden</u> edges.

---

1. Draw dotted lines to show the hidden edges.

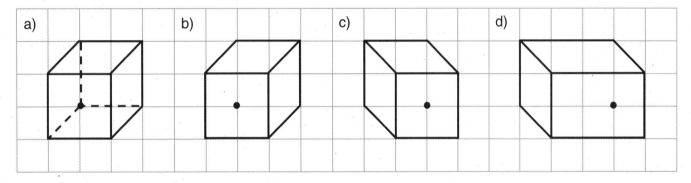

a)   b)   c)   d)

2. Shade all of the edges (the first one is started).
   Count the edges as you shade them.

a)
_____ edges

b)
_____ edges

c)
_____ edges

d)
_____ edges

e)
_____ edges

f)
_____ edges

g)
_____ edges

h)
_____ edges

3. Vertices are the points where the edges of a shape meet.
   Put a dot on each vertex. Count the vertices.

a)
_____ vertices

b)
_____ vertices

c)
_____ vertices

d)
_____ vertices

Geometry 2

4. Shade the...

front face:

a)   b)   c)   d)

back face:

e)   f)   g)   h)

side faces:

i)   j)   k)   l)

top and **bottom** faces:

m)   n)   o)   p)

back face:

q)   r)   s)   t)

bottom face:

u)   v)   w)   x)

5. Shade the edges that would be hidden if the skeleton was covered in paper and placed on a table.

a)   b)   c)   d)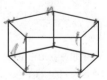

BONUS

6. Shade the edges that would be hidden if the skeleton was covered with paper and was hung above you in the position shown.

G6-36: Prisms and Pyramids

The solid shapes in the figure are called **3-D shapes**.

Faces are the flat surfaces of a shape, **edges** are where
two faces meet, and **vertices** are the points where 3 or more
faces meet.

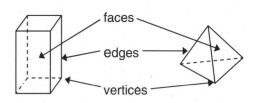

Pyramids have a **point** opposite the base. The base of the shape is a polygon; for instance, a triangle,
a quadrilateral or a square (like the pyramids in Egypt), a pentagon, etc.

Prisms do not have a point. Their bases are the same at both ends of the shape.

----

1. Count the faces of each shape.

a)

_6_ faces

b)

_6_ faces

c)

_6_ faces

d)

_4_ faces

e)

_5_ faces

f)

_9_ faces

g)

_4_ faces

h)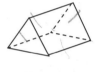

_4_ faces

2. Using a set of 3-D shapes and the chart below as reference, answer the following questions.

A	B	C	D	E
Square Pyramid	**Triangular Pyramid**	**Rectangular Prism**	**Cube**	**Triangular Prism**

a) Describe each shape in terms of its faces, vertices and edges. The first one has been done.

	A	B	C	D	E
Number of Faces	5	4	6	6	5
Number of Vertices	5	4	?		
Number of Edges	8		8	8	6

b) Did any pair of shapes have the same number of faces, vertices or edges? If so, which shapes
share which properties?

Geometry 2

G6-37: Prism and Pyramid Bases

Melissa is exploring differences between pyramids and prisms. She discovers that …

- A **pyramid** has **one base**.
  (There is one exception – in a triangular pyramid, any face is a base.)

  Example: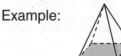

- A **prism** has **two bases**.
  (There is one exception – in a rectangular prism any pair of opposite faces are bases.)

  Example: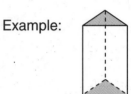

IMPORTANT NOTE:
The base(s) are not always on the "bottom" or "top" of the shape.

- -

TEACHER:
The activity that goes with this worksheet will help your students identify the base of a 3-D figure.

1. Shade the base <u>and</u> circle the point of the following pyramids.
   NOTE: The base will not necessarily be on the "bottom" of the shape (but it is always at the end opposite the point).

   a)   b)   c)   d)

   e)   f)   g)   h)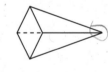

2. Now shade the bases of these prisms.
   REMEMBER: Unless all its faces are rectangles, a <u>prism</u> has <u>two bases</u>.

   a)   b) 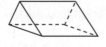   c)   d)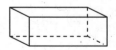

   e)    f)   g)   h)

3. Kira has many prisms and pyramids. Can you circle the ones that have **all congruent faces**?

a)   b) 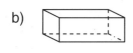   c)    d)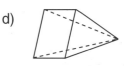

e)   f)   g)   h)

4. Shade the bases of the following figures. Be careful! Some will have two bases (the prisms) and others will have only one (the pyramids).

a) 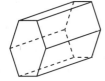   b) 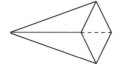   c)    d)

e)   f)   g)   h)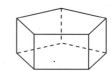

i)   j) 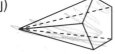   k) 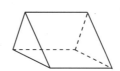   l)

m)   n)   o) 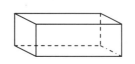   p)

5. "I have a hexagonal base." Name two 3-D shapes this could describe.

1. Circle all the **pyramids**. Put an "X" through all the **prisms**.

2. Match each shape to its name. The first one has been done for you.

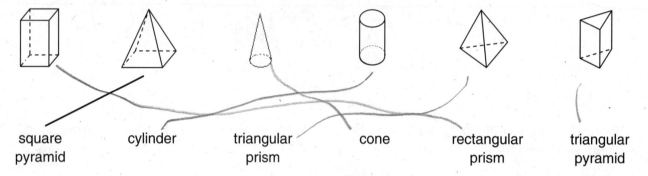

square pyramid   cylinder   triangular prism   cone   rectangular prism   triangular pyramid

3. Use the chart to find properties that are the <u>same</u> and <u>different.</u> Then complete the sentences below.

Property	Triangular Prism	Square Pyramid	Same?	Different?
Number of faces	5	5	✓	
Shape of base	traigle	▭		✓
Number of bases	2	1		✓
Number of faces that are <u>not</u> bases	3	4		✓
Number of edges	6	5		✓
Number of vertices				

a) A triangular prism and a square pyramid are the <u>same</u> in these ways: _____

_____

_____

b) A triangular prism and a square pyramid are <u>different</u> in these ways: _____

_____

_____

4. a) Complete the following property chart. Use the actual 3-D shapes to help you.

Shape	Name	Number of...			Pictures of Faces * In each case, circle the base(s)
		edges	vertices	faces	
	triangle	4		4	
	rectangle	8		6	
		6		5	
	trange	5		5	

b) Count the number of sides in the base of each pyramid. Compare this number with the number of vertices in each pyramid. What do you notice?

c) Count the number of sides in the base of each prism. Compare this number with the number of vertices in each prism. What do you notice?

5. Compare the sets of shapes below. For both a) and b), <u>name</u> the shapes first, and then write a paragraph outlining how they are the <u>same</u> and how they are <u>different</u>.

a)   b)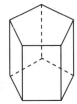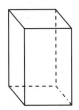

BONUS
6. Draw rough sketches of as many everyday objects you can think of that are (or have parts that are) pyramids or prisms.

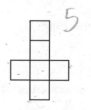

| triangular pyramid | square pyramid | pentagonal pyramid | triangular prism | cube | pentagonal prism |

1.

Name of Figure	Shape of Base	Number of Faces	Number of Edges
1	trangle	4	6
2	▢, △	5	8
3	⬠, △	6	10
4	▢, △	5	10
5	▢	6	8
6	⬠, ▢	7	11

2. Draw the missing face for each net.

i)

ii)

iii)

a) What is the shape of each missing face?

<u>trangle</u>

b) Are these nets of pyramids or of prisms? How do you know?

_____

_____

3. Draw the missing face for each net.

i)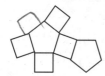

ii)

iii)

a) What is the shape of each missing face?

b) Are these nets of pyramids or of prisms? How do you know?

_____

_____

4. Name the object you could make if you assembled the shapes.

a)

b)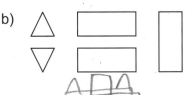

c)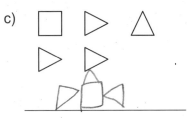

5.

A: B: C: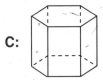

Shade the base of each shape above and then fill in the chart below.

	A	B	C
Number of sides on base	4	5	6
Number of triangular faces	4	5	6

What relationship do you see between the number of sides on the base and the number of triangular faces on the pyramid?

6.

A:    B: C:

Shade the bases of each shape and then complete the chart below.

	A	B	C
Number of sides on base	4	5	6
Number of (non-base) rectangular faces	2	5	0

What relationship do you see between the number of sides on the base and the number of (non-base) rectangular faces on the prism?

7. How many of each type of face would you need to make the desired 3-D shape?

a)

△ = 4

▭ = 1

b)

▭ = 2

▭ = 4

c)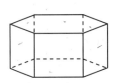

▭ = 6

⬡ = 2

Geometry 2

1. Evelyn sorts the following figures using a Venn diagram. She first decides on two properties that a figure might have and makes a chart (see below). She then writes the letters into the chart and checks to see which figures share both properties.

A B C D E F

a)

Property	Figures with this property
1. One or more triangular faces	A, C
2. Six or more vertices	

b) Which figure(s) share both properties? _____

c) Complete the following Venn diagram.

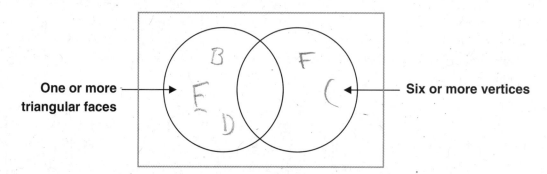

2. Complete both the chart and the Venn diagram below using the shapes A to F in Question 1.

a)

Property	Figures with this property
1. Rectangular base	C E
2. Ten or more edges	F E

b) Which figures share both properties? _____

c) Using the information in the chart above, complete the following Venn diagram.

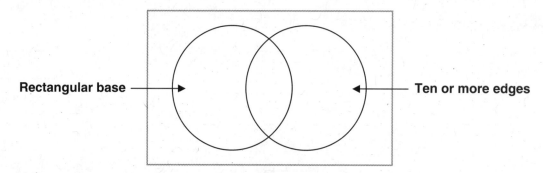

3. Pick a pair of properties and draw a Venn diagram to sort these shapes.

G6-41: Creating Patterns with Transformations

1. Rotate each figure around the point P.

a)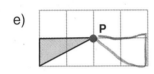

90° clockwise

b)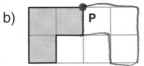

90° counter
clockwise

c)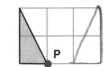

90° clockwise

d)

90° clockwise

e)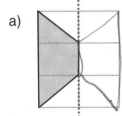

180° counter
clockwise

f)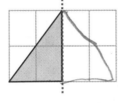

90° clockwise

g)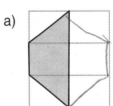

180° clockwise

h)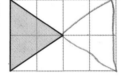

90° clockwise

2. Reflect each figure in the mirror line.

a)

M

b)

3. Slide each figure 1 unit right.

a)

b)

4. Extend each pattern. Then describe the <u>two</u> transformations used to create the pattern. Draw in any mirror lines or points of rotation.

a)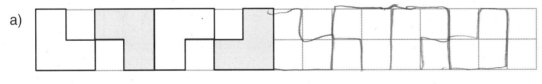

Description: _____

b)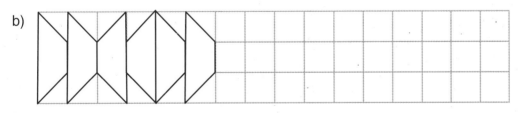

Description: _____

c)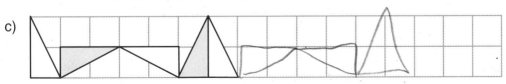

Description: _____

jump math
MULTIPLYING POTENTIAL

Geometry 2

5. Trace and cut out the shape below. Make a pattern by …

    a) Sliding the shape repeatedly one unit right.

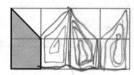

    b) Reflecting the shape repeatedly in the mirror lines.

    c) Rotating the shape repeatedly 180° around the dots.

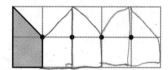

6. Each of the patterns below was made by repeating a transformation or a combination of transformations. Use the words "slide", "rotation" or "reflection" to describe how the shape moves from …

(i) Position 1 to 2   (ii) Position 2 to 3   (iii) Position 3 to 4   (iv) Position 4 to 5

a)

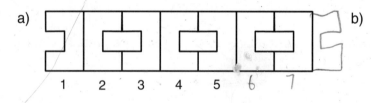

b)

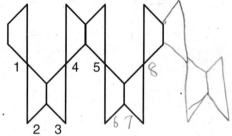

c)

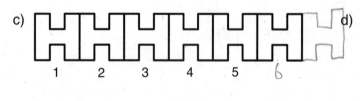

d)

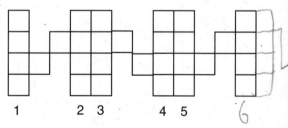

    e) Choose a pattern above that can be described in two different ways. Which two single transformations could each produce the pattern?

7. Draw a shape on grid paper and make your own pattern by a combination of slides, rotations and reflections. Explain which transformations you used in your pattern. I used cubes

G6-42: Isoparametric Drawings

Follow these steps to draw a **cube** on isometric dot paper:

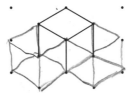

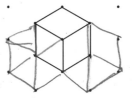

Step 1:
Draw the top square with
4 vertices at 4 different dots.

Step 2:
Draw vertical lines at 3 vertices
to touch the dots below.

Step 3:
Join the vertices.

- -

1. Draw the following figures constructed with the interlocking cubes on isometric dot paper. The first one has been started for you.

a)

b)

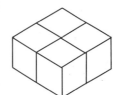 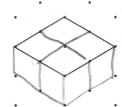

c)

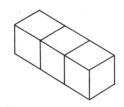 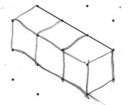

d)

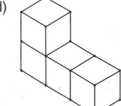 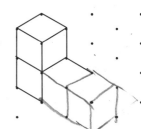

e)

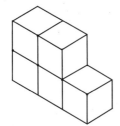 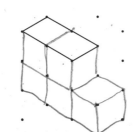

f)

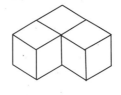 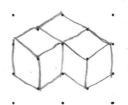

BONUS

2. Draw the following figures constructed with interlocking cubes on isometric dot paper.

a)

b)

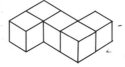

c)

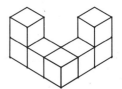

d)

e)

f)

G6-43: Drawing Figures and Mat Plans

1. Build with blocks or interlocking cubes.

a)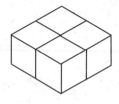

b)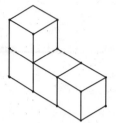

c)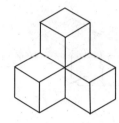

d)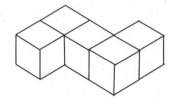

e)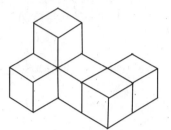

f)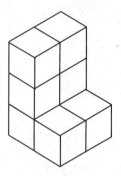

2. Fill in the numbers in the "mat plan", then build the figure. The first one was done for you.

a)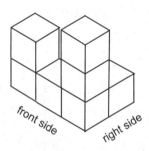

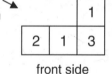

b)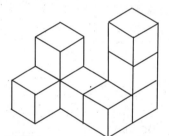

c)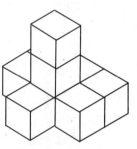

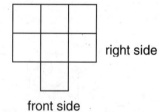

3. Draw a "mat plan" (as in Question 2), then build the figure.
   HINT: Shade the squares facing the front side.

a)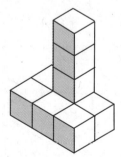

b)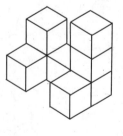

c)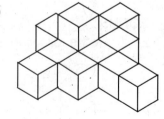

4. Build three figures with 12 cubes. Draw a mat plan for each.

Geometry 2

G6-44: Orthographic Drawings

1. Given a structure made with cubes, you can draw a front, top and side view as shown.

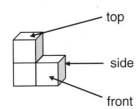 top / side / front

front view

side view

top view

Draw a front, top and side view for the following structures.
HINT: Use cubes to help you.

a)   b) 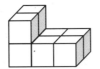   c) 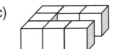   d)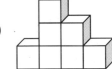

2. a) Build a structure with these views.

(i)
top view   **front view**   **side view**

(ii)
top view   **front view**   **side view**

b) Draw a mat plan for both figures.

3. Which picture could be the right-side view of this structure? Circle it.

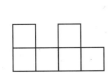

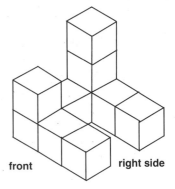
front   **right side**

4. Using centicubes, build two different shapes that have a volume of exactly 10 cubic centimetres. Draw a mat plan of each of your shapes.

5. How many different rectangular prisms can you build with 8 cubes? Draw a mat plan for each of your shapes.

6. Use grid paper to draw orthographic (front, top and side) views of the structures given by the mat plans.

a)   b)   c)   d)

a) 1 / 1 1 3
b) 2 1 / 1
c) 2 1 / 2 / 2
d) 1 1 3 / 1 / 2

jump math
MULTIPLYING POTENTIAL.

Geometry 2

1.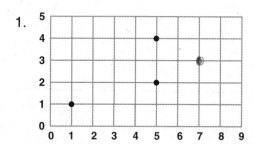

   a) Add a dot so that the four dots form the vertices of a parallelogram.

   b) Write the coordinates of the four vertices.

   (1 , 1) (5 , 2) (5 , 4) (7 , 3)

2. Sketch what each letter would look like reflected in the mirror line.

   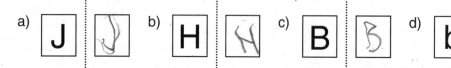

   a) J   b) H   c) B   d) b   e) e

   Find 5 letters of the alphabet that look the __same__ after a reflection.

   f) b   g) B   h) e   i) H   j) J

   Then find 5 that look __different__.

   k)   l)   m)   n)   o)

3. Match the description of the figure with its name.

   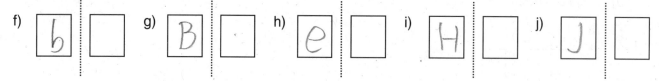

   _____ cone   **A.** I have 6 congruent faces.

   _____ triangular prism   **B.** I have 5 faces: 2 triangles and 3 rectangles.

   _____ cube   **C.** I have 4 faces. Each face is a triangle.

   _____ cylinder   **D.** I have 2 circular bases and a curved face.

   _____ triangular pyramid   **E.** I have 1 circular base and a curved face.

4. List any 3-D shapes that have each of these properties:

   a) "I have 5 faces."   b) "I have 12 edges."   c) "I have 6 vertices."

5. What is similar about a triangular prism and a triangular pyramid? What is different?

6.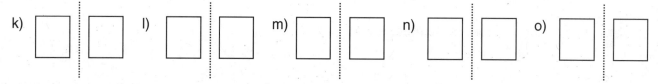

   a) Describe a transformation that would leave the arrow pointing in the same direction.

   b) Describe a transformation that would leave the arrow pointing in the opposite direction.

   c) Describe a transformation that would leave the arrow pointing at a right angle to the original.

To learn about the full range of JUMP Math publications, including pricing, discount, and ordering information, visit **www.jumpmath.org**.

If you question whether some students are capable of learning math, you will find John Mighton's books enlightening.

The Myth of Ability
Nurturing Mathematical Talent in Every Child

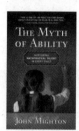

Mighton tells JUMP Math's fascinating story and explains its teaching method. He also provides lots of simple problems to get students started on the excitement of mastering math. Equal parts inspiration and instrument, this book shows how Mighton's empowering program gives children the tools they need to love learning. *The Myth of Ability* will transform the way you look at math education.

ISBN: 978-0-88784-767-7
Published by House of Anansi Press

The End of Ignorance
Multiplying Our Human Potential

The End of Ignorance conceives of a world in which no child is left behind—a world based on the assumption that each child has the potential to be successful in every subject. Mighton challenges us to re-examine the assumptions underlying current educational theory. He pays attention to how students pay attention, chronicles what captures their imaginations, and explains why their sense of self-confidence and ability to focus are as important to their academic success as the content of their lessons.

ISBN: 978-0-676-97962-6
Published by Knopf Canada

John Mighton is the founder of JUMP Math (Junior Undiscovered Math Prodigies), an educational charity providing community and educational outreach, and professional training. He won an NSERC fellowship in mathematics at the Fields Institute, and has taught at McMaster and the University of Toronto. He is an Ashoka Fellow, and an award-winning playwright.